ADVANCE PRAISE FOR *THE WONDERFUL FUTURE THAT NEVER WAS*

"A wonderful coffee-table book that you will also want to read page by page, this book combines a tongue-in-cheek survey of the most outrageous, wrong predictions of future wonders with some amazingly good guesses. 'The dance between society and its engineers,' the authors say, 'steps both backwards and forwards, a foxtrot of progress.'"

—Joe Haldeman, Hugo and Nebula-winning author of *The Forever War*

"A beautifully illustrated odyssey into the minds of a generation who believed that science had an answer to every question—even the ones that nobody asked. We may not have the flying car yet, but who can complain? Reality could never compare to these vivid dreams spun during the golden age of science fiction."

—Daniel H. Wilson, Ph.D., author of *Where's My Jetpack?*

"All these fantastically fabulous futures, and I get to live in *none* of them, and no, having an iPod Touch does *not* make up for it. But at least I have this book. Which almost *does* make up for it."

—John Scalzi, bestselling author of *Old Man's War*

"It's fun to mock the impractical devices that never came to be. Benford and *Popular Mechanics* remind us how much of our past's speculations actually occurred—and how they came to influence our path towards the 'future' of today, despite their failures."

—Tobias S. Buckell, NYT bestselling author of *Halo: The Cole Protocol*

"Gregory Benford and the editors of *Popular Mechanics* have provided a fascinating, colorful portfolio of the myriad predictions of past generations, displaying both the vexing difficulties of making accurate prophecies and a charming focus on hoped-for improvements in our everyday lives that is often absent in the extravagant adventures of science fiction. Their book is a compelling, thought-provoking, and just plain entertaining read for people of all ages."

—Dr. Gary Westfahl, critic and Hugo-nominated editor of *Science Fiction Quotations*

THE WONDERFUL FUTURE THAT NEVER WAS

GREGORY BENFORD
and THE EDITORS of
Popular Mechanics

HEARST BOOKS
New York

CONTENTS

FOREWORD 6

INTRODUCTION 8

CHAPTER 1: THE CITY OF THE FUTURE 16

CHAPTER 2: HOME, SWEET HOME OF TOMORROW 36

CHAPTER 3: MIND & WORD BECOME FAR-REACHING & UNIVERSAL 76

CHAPTER 4: HEAVY WATER MAY PROLONG LIFE 102

CHAPTER 5: AIRSHIPS SUPERSEDE BATTLESHIPS 120

CHAPTER 6: THIS UNFINISHED WORLD 168

INDEX OF PREDICTIONS 203

INDEX 205

When *Popular Mechanics* magazine launched in 1902, it seemed almost impossible for our editors and writers to imagine what 2010 might look like, but that didn't stop them from trying. Since then, we've published countless predictions of scientists, inventors, and other visionaries as steam gave way to electricity, stone buildings were overshadowed by skyscrapers of concrete and steel, and advances in transportation and telecommunications seemed to shrink the world.

Now we live in an era we're used to thinking of as "the future." Though we still lack flying cars and jet packs (as predicted in 1928 and 1964), our clothes are made of milk (as we forecast in 1929), our foods are fortified with grass (1940), and our mail is sorted by robots and delivered by airplane (though perhaps not the way we anticipated in 1921—see page 83). Surrounded by wonders and a fast-evolving culture of innovation, it's just as challenging today for us to imagine the next century as it must have been for our early 20th century colleagues to envision the fabled year 2000.

So we decided it was high time to take a look back at the predictions of the past, not only to score them on accuracy (some are shockingly prescient, some hilariously wrong) but also to pay tribute to the inventiveness of the past. We hope this collection of articles and essays, which first appeared in *Popular Mechanics* between 1903 and 1969, will inspire fabulous new visions as well as recalling the sense of wonder that previous generations of readers felt when they craned their necks at the first skyscrapers, read about the first heart transplant, and watched the first man walk on the moon. Power up your personal helicopter and join us on a glorious adventure in the many wonderful worlds of tomorrow.

The Editors of *Popular Mechanics*

PREDICTION
1932

There are those who feel our present difficulties are due to the problems which rapid progress has raised, but such developments are the sign posts of advancing civilization. From the laboratory will come the technical solutions to the problems that technology has created; achievement will find the way, and education will light the road to further progress. Will man continue to be fit to live in the new universe his brain is creating, or will he be crushed by his Frankenstein? It is self-evident that our souls must grow with science or die by science. We think and hope that man, who has been made by his tools, will continue to be their master.

BY GREGORY BENFORD

AFTER THE FUTURE—WHAT?

I've often noted that within the first thirty seconds of a conversation about the future, the past usually comes up to support a prediction. All our knowledge is about the past, but all our decisions are about the future. The past helps us to think forward and to learn from our mistakes and successes. The idea that if it was true yesterday then it will be true tomorrow drives many of our decisions.

The hardest questions about that future do not concern what we know (at least are familiar) about—"Will I get that job offer?" Rather, they concern events that come at us as surprises: "Who thought of *that*?" So we try to see forward, through a fog of uncertainty.

One good way to answer such questions is to look back at our dreams as depicted in musty old magazines, and learn from what looked like pretty good bets to our forefathers. Inspecting the past is both amusing and instructive. That's what this book is about—learning from our nostalgia.

In the yellowing pages of magazines like *Popular Mechanics* you encounter such grandiose headlines as

- We're About to Lick Hurricanes
- A House Blooms like a Flower
- Airport in the Heart of a City Provided by Logical Design
- Nonstop Airplane Mail Delivery by Parachute
- The Great Wall of China to be Motor Highway
- Take a Long Weekend—on the Moon

Such sunny views hold more than nostalgic interest. Often they were right, and grasped at least a portion of the future. The illustrations are often more interesting than the articles themselves, because they contain details that seemed logical at the time but bring howls of laughter now. Fashion seems especially static in that past future: women fly in jets wearing ankle-length skirts with bustles, and seated men keep on their hats.

In the year 1900 everyone knew that technology drove their world and would drive the future even harder. That was the single most prescient "prediction" of the twentieth century; not only was it true, it was self-fulfilling.

Granted, *Popular Mechanics* probably wasn't the orthodox view from the 1930s. It and other such magazines spoke to those who naturally thought of technically enhanced futures. But was *Popular Mechanics* so different from today's "major new breakthrough!" articles in our newspapers?

Predictions might be more restrained and more subtle today, but that doesn't mean they're better or more accurate. They're just more recent, so we think they're hip, knowing, aware. Beware our prejudice in favor of what's recent—history doesn't only repeat itself, it sometimes stutters.

Predictions by optimists held up better than the gloom of pessimists throughout the twentieth century—despite the two big World Wars. Throughout that age, it was smart to follow the famous saying by a man who was arguably the most influential figure of that time: "Imagination is more important than knowledge." That's Albert Einstein speaking.

WASN'T THE FUTURE WONDERFUL?

***There are two futures, the future of desire and the future of fate, and man's reason has never learnt to separate them.* —J. D. Bernal, 1929**

Americans were not always lovers of technology. Thomas Jefferson distrusted manufacturing generally, and in 1829 a highly regarded governor of New York wrote to the president:

As you may well know, Mr. President, "railroad" carriages are pulled at the enormous speed of fifteen miles per hour by "engines" which, in addition to endangering life and limb of passengers, roar and snort their way through the countryside, setting fire to crops, scaring the livestock and frightening women and children. The Almighty certainly never intended that people should travel at such breakneck speed.

The governor was Martin Van Buren, who became the eighth President of the United States from 1837 to 1841.

But as a new century dawned, attitudes changed. Marconi, Edison, and the Wright brothers made wonders seem to come at a faster and faster pace. Excitement gathered, and American was in the lead.

In this spirit, the visionary forward-thinkers of the twentieth century nailed many things that did come to pass, like television and freeways. Often, though, society got to those end results along very different paths, and with far different consequences, than the technophiles had predicted. A classic example is the rise of the automobile. Many saw that it would give us new freedom of movement, and so envisioned cities with broad avenues and freeways—as happened. But few foresaw the social consequences of the commuter society: dull suburbs and the rise of fast-moving criminals like John Dillinger and Bonnie and Clyde. No one at all predicted that sexual mores would change, as teenagers got away from their parents and into the back seats of cars.

Dreams of the future about large-scale engineering went especially awry. Gargantuan schemes got proposed without thinking

about who would want such things. (If it looks cool, do it!) Consider the images of huge domes over cities to ensure clean air, a skyscraper dwarfing the Eiffel Tower and shaped like a fighter jet, six-level highways, cities under the sea. Why don't we see these today? Studying our past dreams then becomes an exercise in how and why some things look good, sound great—but just aren't practical.

Bernal's "future of fate" is the real future, and it seldom coincides with our futures of desire. The twentieth century was a vast laboratory that tested our expectations against the unfolding reality. How well did that experiment perform?

This book looks fondly back at how wonderful our dreams were and asks questions that are useful for us now:

- **Which predictions actually happened?**
- **Which didn't?**
- **What can we learn from this, to sharpen our own predictions?**

I'll draw some conclusions as this book unfolds.

Radio, the telephone, the airplane, the automobile, television, laptop computers:

PREDICTION 1912

A German motor-sleigh of unusual design travels at a speed of 60 miles an hour. An automobile motor occupies the center of the body and drives an aerial propeller. Steering is accomplished by means of the light forward set of runners.

all of them are familiar now, so ordinary that we forget how drastically these once-wondrous technologies seemed. Yet they altered everything.

Go back to the early nineteenth century and try to get across England. You would have to go via stagecoach, the only method possible; it would have taken several days and cost about a month's wages. Plus, it was a jouncing, uncomfortable ride. In 2010 you pay a weeks' wages to fly in comfort to the other side of the planet.

By the year 1900 we had rail and simple automobiles. Gradually, transport improved. Anyone in 1910 would have been dazzled by flying in an aluminum tube at a high fraction of Mach 1—and listening to music, watching movies, or sipping wine all the while. Yet that everyday wonder we take for granted—and giant rockets like the Saturn V—were implicit in the Wright Flyer. That humble airplane made using giant spruce wood flew shorter than the wingspan of a Boeing 747 on its first flight. Its greatness lay in its promise.

Similarly, the iPod lay implicitly in the hand-cranked Victrola that spun odd black disks of wax. The Internet was a dream lying dormant in the telegraph (invented in the 1830s) and indeed, in the ordinary rotary dial phone on the wall. Radio implied the TV that had already replaced the piano in the parlor. To a great extent, our modern routine wonders emerged and evolved from ideas centuries old, but did so in an accelerating age of wonder.

There were plenty of critics. Nobelist Ernest Rutherford, who proved that the atomic nucleus existed, nonetheless said any actual use of nuclear energy was impossible. A royal astronomer said in the 1940s that talk of going to the moon was "bilge." When Boeing's first airplane that could carry ten people rolled out, one of the company's engineers said, "There will never be a bigger plane built." Bell Telephone projected that as many as a million people would eventually have their own private lines. An IBM executive imagined that as many as a hundred businesses would eventually use computers—which would fill whole rooms.

Against naive optimism, some magazines carried articles claiming "most scientific fiction can't come true." Today's editor of MIT's *Technology Review,* Jason Pontin, has said wryly, "Science fiction is to technology as romance novels are to marriage: a form of propaganda." This is true, but no one can accomplish anything without first imagining it.

Throughout the last century, many made fun of wacko schemes such as teleportation booths, tunnels beneath the English Channel, travel to the moon, and radio communications with Mars. I wonder if they'll be as right about teleportation as those other claims?

Some decry fiction and even *Popular Mechanics* as having any true predictive power. (Science fiction, they say, is really about the present day.) Nonetheless, the accurate predictions of many writers are justly famous. Geostationary telecommunications satellites were first proposed by Arthur C. Clarke in a paper titled "Extra-Terrestrial Relays: Can Rocket Stations Give World-Wide Radio Coverage?" published in *Wireless World* in October 1945. Space travel has been a staple of science fiction since Jules Verne published *De la Terre à la Lune* in 1865. Robots first appeared in Karel Capek's play *R.U.R.* in 1921. Indeed, it is more useful to ask, what *hasn't* fiction and smart projection predicted?

Today, many writers are now less interested in predicting and thus determining the future precisely because they do not believe that linear, programmatic determinism is the right angle of attack. They see themselves more as conceptual gardeners, planting for fruitful growth, rather than engineers designing eternal, gray social machines.

One of the twentieth century's great visionaries, Freeman Dyson, said, "Science is my territory, but science fiction is the landscape of my dreams." I agree. I like technologies that expand our sense of what it might mean to be human. Many of us came to technology through science fiction; our imaginations remain secretly moved by science-fictional ideas.

Discerning a causal relationship between what science fiction has predicted and what technologists have created might be an instance of the logical fallacy *post hoc ergo propter hoc* ("after this, therefore because of this"), except for one curious fact: science fiction writers and technovisionaries not only describe current research and extrapolate its likely development but also prescribe cool things that enthrall technologists later make or try to make. In short, life imitates art.

We cannot have a future that we do not first imagine. Historians often convey the impression that the past, since it is now fixed, was a neat, cut-and-dry time. This mistake makes the present seem messy. The past is a far country, but the distance should not confuse us about its turbulent nature. This book shows some of the brilliant projections and clear misses of the twentieth century. The twenty-first will be similarly confused.

We shape our future with incomplete information, then must live with what results. Projecting technologies helps us consider upsides and downsides, risks and opportunities. Americans usually side with opportunity over risk, and so have led the world in technology for over a century.

This book aims to give you some surprises, some rueful laughs, and a sense of technical perspective. Science fiction writers might well use it as a guide to non-silly predictions. Artists may acquire a sense of realistic craziness. It may stimulate some to what the entrepreneur Alex Lightman calls "blueprint prophecy"—not only showing what may happen but also conveying the promise: "If you build it, they will come. And pay you. And think you are cool for making dreams come true."

If this book merely jogs those little gray cells or gives a good time, it will have done its job.

THE FUTURE FACTORY

"The future isn't what it used to be."

—Arthur C. Clarke

The past has a vote, but not a veto. Looking at decades-old projections, one notices the old-fashioned lens through which the past saw inventions we now use routinely—computers, jet travel, the Internet. Yesterday's visions of tomorrow often understood the new gizmo, but missed the new context.

Even when imagination is oozing marvels not yet invented, the audience of that era sees them in their context, not ours. Now, though, we notice ladies with vast hats and men with spats in illustrations of jetliners. They're shown gazing in wonder at passing landscapes, not watching in-flight movies or writing e-mail. Does this matter?

No. Imagination shapes the future but does not command it. The questions I posed above—*Which predictions actually happened? Which didn't? What can we learn from this to sharpen our own predictions?*—are themselves attempts to use the past to better form the future.

I found the percentage of good predictions in these magazines surprisingly high: better than 50 percent. Failures usually assumed that bigger would always be better—vast domed cities, floating airports, personal helicopters, tunnels across continents. Ideas also fail because adding functions compromises performance (the aircars that were actually built in the 1950s were inferior cars and lousy planes).

Optimism paid off, though. The most infamous attempt to predict the socioeconomic future was the Club of Rome's *The Limits to Growth*. In 1972 it foresaw only dwindling resources, allowing for no substitutions or innovation. A famous bet over the price of metals in 1990 led to the club's public debacle; copper was cheaper later, not a fought-over commodity, as predicted. The 1970s oil shocks lent their work credence, but luckily, markets have erased the gloomy, narrow view of how dynamic economies respond to change.

Rather than looking at the short run and getting it wrong because we project linearly, we should consider peering over the heads of the mob's immediate concerns. That means tracing long-run ideas that do not necessarily parallel the present. A historical example of this is J. D. Bernal's *The World, the Flesh and the Devil*, which examined our long-term prospects in terms that seemed bizarre in 1929 but resonate strongly today: engineered human reproduction, biotech, our extension into totally new environments such as the deep oceans and outer space.

Paging through musty magazines, far from the electronic herd, I felt the future envy of past eras. Americans are great optimists, willing to get hooked on a vision—whether Utopian or just plain cooler than today—and we want to bring it into being.

The most profound idea these works can give us comes when a teenager's eyes widen and you hear: "I want to solve that problem," or "I want to live in that world!"

In the end, the best way to predict the future is to create it. ◙

1
THE CITY OF THE
FUTURE

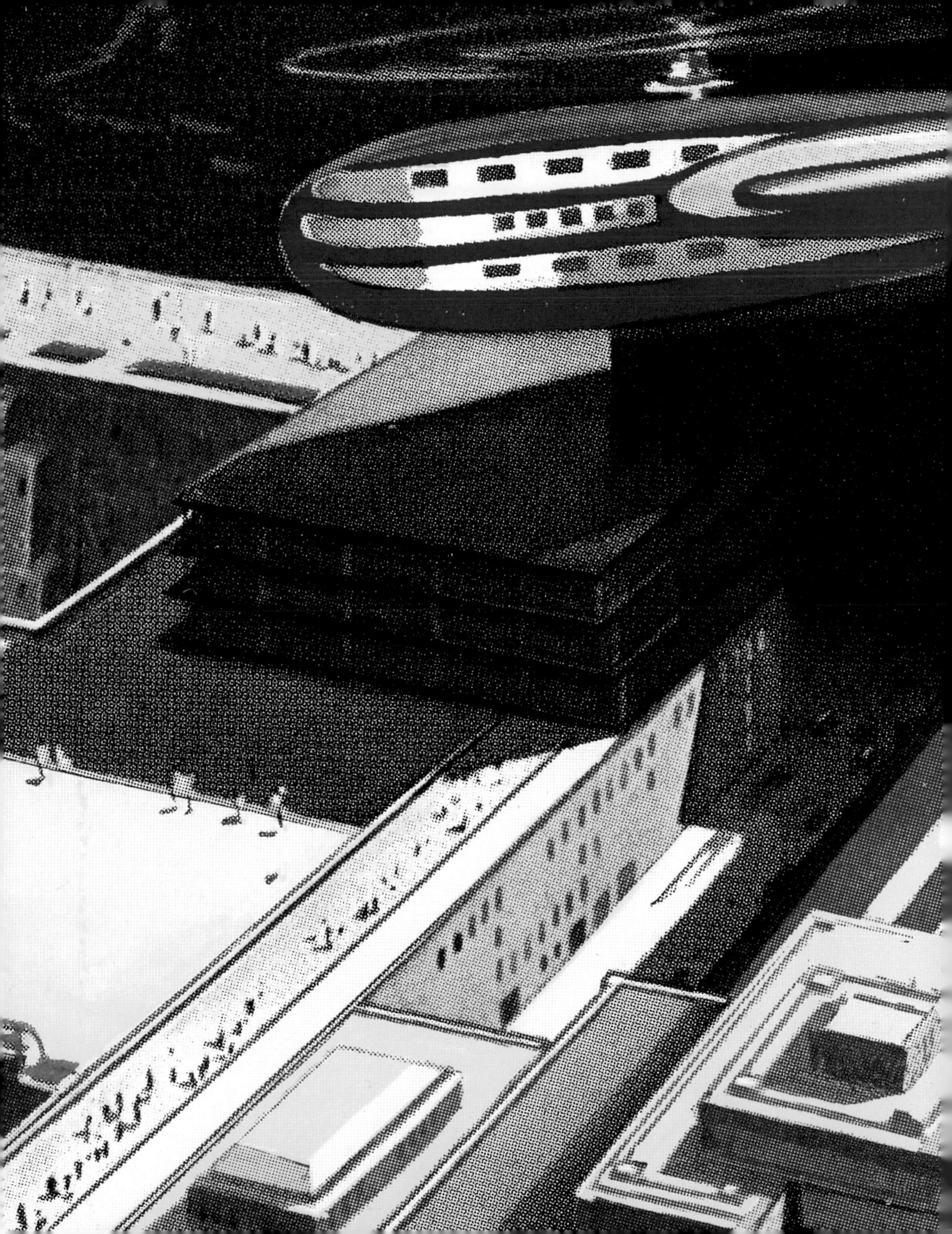

For cities, the future promised progress, because by the early 1900s the rural way of life was definitely a thing of the past.

Twentieth-century Americans got visions of better homes and streets from world's fairs and visual lures like the megastructures of Buckminster Fuller. Fuller's skyscrapers were enormous hyperboloid towers—"machines for living" a hundred stories high. (Freed of gravity, the same grand scales turned up in space stations.) These visions were secular Edens, without the anti–free market bias of the 1930s or the isolation of the feminist utopias of the 1970s. Their predictions reveal the culture that shaped them.

Zealots abounded. Science fiction writers, auto designers, architects, and advertising mavens—all thought that

PREDICTION 1928

A Venetian-like plan submitted for Chicago solves transportation problems and allows for a pleasant day of shopping.

Among the modernistic buildings proposed by architects is that of a revolving restaurant mounted on a huge column. This affords diners an opportunity for sight-seeing while dining or strolling on platforms.

PREDICTION 1930

PREDICTION 1924

This artist's grouping of recently completed and proposed skyscrapers shows the tendency of modern architecture towards the receding, terraced design of building construction.

technologies could solve social problems better than lectures or spiritual appeals. Futurism, an art movement based on a passionate loathing of everything old, merged with commercial consumerism, reinforcing each other. In the United States, which invented consumer culture, planners thought they could fashion better communities with better buildings.

This fit with older storylines. Americans often saw cities as fabulous yet also dark. To frame this tension, they invented film noir in the 1930s and 1940s, using stylish techniques of shadow and light they borrowed from European directors, especially Fritz Lang. This reuse of older images extended to everything. Even spaceship heroes, from Buck Rogers to Han Solo, fitted out with zippy ships and powerful weapons, came to us straight from the virile mavericks of the Wild West.

Such futures resurrected the past, allowing audiences to buy into technologies that followed well-trodden narrative paths. This helped sell technologies, because if they changed only the material world—leaving social arrangements intact—these futures were easier to digest. So we saw megacities whose men wore bowler hats and attended genteel concerts with ladies whose skirts grazed the ground. Nobody foresaw massive rock-concert amphitheatres with mosh pits filled by skimpily clad sybarites.

Historically, though, technologies hasten social change. Cities change people, and people change cities in response. Science fiction and technologists did not merely sell futures, they shaped them. Fritz Lang's 1926 film Metropolis was hugely influential, but it is based on his 1924 visit to New York, a city that impressed him with its fresh buildings and furious energy. Lang contrasted stunning landscapes of skyscrapers (surely one of the ugliest invented words of the century) with the unseen city. In the shadowy basements cowered oppressed workers enslaved to machines. H. G. Wells foresaw the same imagery in The Time Machine in 1895.

Their agendas got stripped away, though. Audiences carried away images of the sparkling towers; the light, airy suspension bridges; and the swarms of strange vehicles in the sky. They wanted that, and forgot about the basements.

Clearly, cities are winning in the twenty-first century. Megalopolises housing tens of millions grow around the world. In 1900 fewer than one in ten people lived in cities; today, cities house a majority of humanity. Cities are energy efficient compared with suburbs, but can still damage the environment around them. The flight from cities to suburbs, which seems to be waning now, could reverse. What part of the landscape people will ultimately saturate is one of the biggest social questions ahead. ◎

ELEVATED SIDEWALKS AND SUNKEN STREETS

PREDICTION 1923 The builders and architects of today hope to see many dreams realized in the city of the future. Overhead thoroughfares supported on great arches between towering skyscrapers, double-decked streets, moving sidewalks bordered by show windows, and stores connected by artistic bridges and covered promenades will all serve the lives of the citizens.

Airships will land at lofty stations and will carry the businessman and the shopper to the downtown district, where they will be lowered to the streets by speedy elevators. Radio equipment in every office and home will keep the inhabitants advised of the daily happenings in distant parts of the world. A sea of light, radiating from powerful hidden bulbs, will make the night as bright as day.

PREDICTION 1924 Architects and builders have turned their attention to designing gracefully tapered, stepped-in, or pyramidal structures to insure greater access to air and sunlight. The future office building will rise majestically in tiers. Elevated sidewalks and sunken streets will relieve traffic congestion and permits show-window space in second-floor displays. Triple-decked streets divert vehicle, foot and street-railway traffic into three separate

PREDICTION 1928 This block of buildings rises in terraced masses into the skies, and is topped by airplane beacon lights to guide passing flyers.

PREDICTION

1907

GREATER HEIGHTS FOR FUTURE SKYSCRAPERS

Less than a dozen years ago a 20-story building was a world wonder. Now it is insignificant. "The 100-story building is sure to come," says one New York architect.

PREDICTION
1928

Cross section of the future city, with many traffic levels underground; top, street for pedestrians only, and airplane landing fields, above.

Elevated sidewalks may connect various levels of blocks or more or less uniform skyscrapers, while the lofty bridges will be carried over the tower tops.

PREDICTION 1928

PREDICTION
1928

Looking down on the Chanin Building in New York City, from the top of the Chrysler Tower, which is higher than the Woolworth Building.

levels to gain greater freedom for all classes of conveyances and pedestrians.

The familiar skyscraper with its straight severe wall will be an artifact of the past, with the new terraced-block building taking its place. Great heights are expected for future structures; where formerly twenty or thirty stories were previously considered quite tall, now massive buildings of forty to fifty stories, able to accommodate the population of a good-sized city, are planned.

THE SUPER-SUPER-SUPER-SKYSCRAPER

PREDICTION 1913 The future skyscraper will have a climate of its own; its heating, lighting, and ventilating machinery will keep it at a constant temperature. Since the building itself is fireproof, constructed from 4 to 8 inches of vitrified clay or concrete, in its interior wooden finishings and furniture will soon become obsolete.

PREDICTION 1924 Reinforced concrete buildings, 100 stories high, towering 1,000 feet into the air, may soon be seen. Before such a feat can be accomplished, engineers must overcome the impossibility of providing elevator service. In today's 30-story buildings, the weight of the cables supporting the cars is enormous and constructing elevators for buildings over 40 stories is a challenge.

PREDICTION 1928 "The step-back skyscrapers of the future will have moving stairs on the outside of the buildings instead of elevators, with facilities for passengers to alight at any floor," declared Harvey Wiley Corbett, noted city planner. "Some of the skyscrapers will be a half mile high and will house small-sized cities."

Major Henry H. Curran of New York, opposes Mr. Corbett and points to the impossibilities of such buildings in terms of human happiness as well as construction, saying, "There are everyday workers who count their ribs on release from the elevators and subways that take them to their offices. They must pop out of kiosks like prairie dogs."

TRUE!

The tallest building in today's world is Burj Khalifa in Khalifa. At 2,684 feet, it's just over a half mile high, although its use is primarily commercial and it lacks the moving stairs.

PREDICTION 1932 In a few years, perhaps even in a few months, we may expect skyscrapers to go up in 180 days and be replaced in two decades. These structures will have their inception in the laboratory rather than the stone quarry and the lumber pile, for these skyscrapers will soon be composed almost entirely of synthetic materials.

THE END OF THE AGE OF STEEL

PREDICTION 1927 From the stone age, civilization has passed through various ages to today's age of steel. Now, another substance threatening metal's position: glass, one of the most fragile of all and, until recently, used chiefly for decoration and illumination. Modern research has developed new qualities in glass, some with enormous heat-resisting powers, and another grade so hard it can be struck tremendous blows with a hammer and cannot be broken.

PREDICTION 1933 From London comes the prediction of reinforced plastics replacing metal for automobile bodies, airplane fuselages, yacht hulls, and buildings. The strength of plastic products is comparable to that of iron and steel, according to H.B. Potter, of the Society of Chemical Industries. "Cement in itself was of little use to builders as a main structural material until it was reinforced," says Mr. Potter. "By

A woman stretches plastic filler for safety glass.

PREDICTION 1936 The photograph, above, made in a darkened room, the only light coming from an adjoining room, separated by a wall of glass blocks. The blocks diffuse the light rays.

reinforcing plastics we look forward to materials not only as strong, but considerably lighter than those now in use. I look forward to molding walls of homes which will be fireproof, scratchproof, and unstainable."

PREDICTION 1936 Visualize a city of shining glass. You walk to work on glass pavements, enter a skyscraper of colored glass, are whisked upward in a glass-enclosed elevator shaft, enter an office with glass floors, walls and ceiling—and perhaps doff a topcoat of fibrous glass before sitting down at a glass desk.

Glass today plays an increasingly important part in architecture, both as a structural and ornamental material, as a result of the development of glass bricks, tiles and other building products. Some architects see the city of the future composed of steel or aluminum framed structures with walls, floors and ceilings entirely or chiefly of glass.

As a building material, glass offers the hardness of steel, the durability of granite or marble and, in addition, it provides a new medium enabling the architect to use walls, ceilings, and even floors in his lighting scheme.

PREDICTION 1957

SHELTERS OF *SPRAYED PLASTIC*

Home builders of the future may be able to erect a house in less than a day, using only steel tubing, cloth tape and a quick-drying plastic spray. The technique was first used to mothball Navy ships.

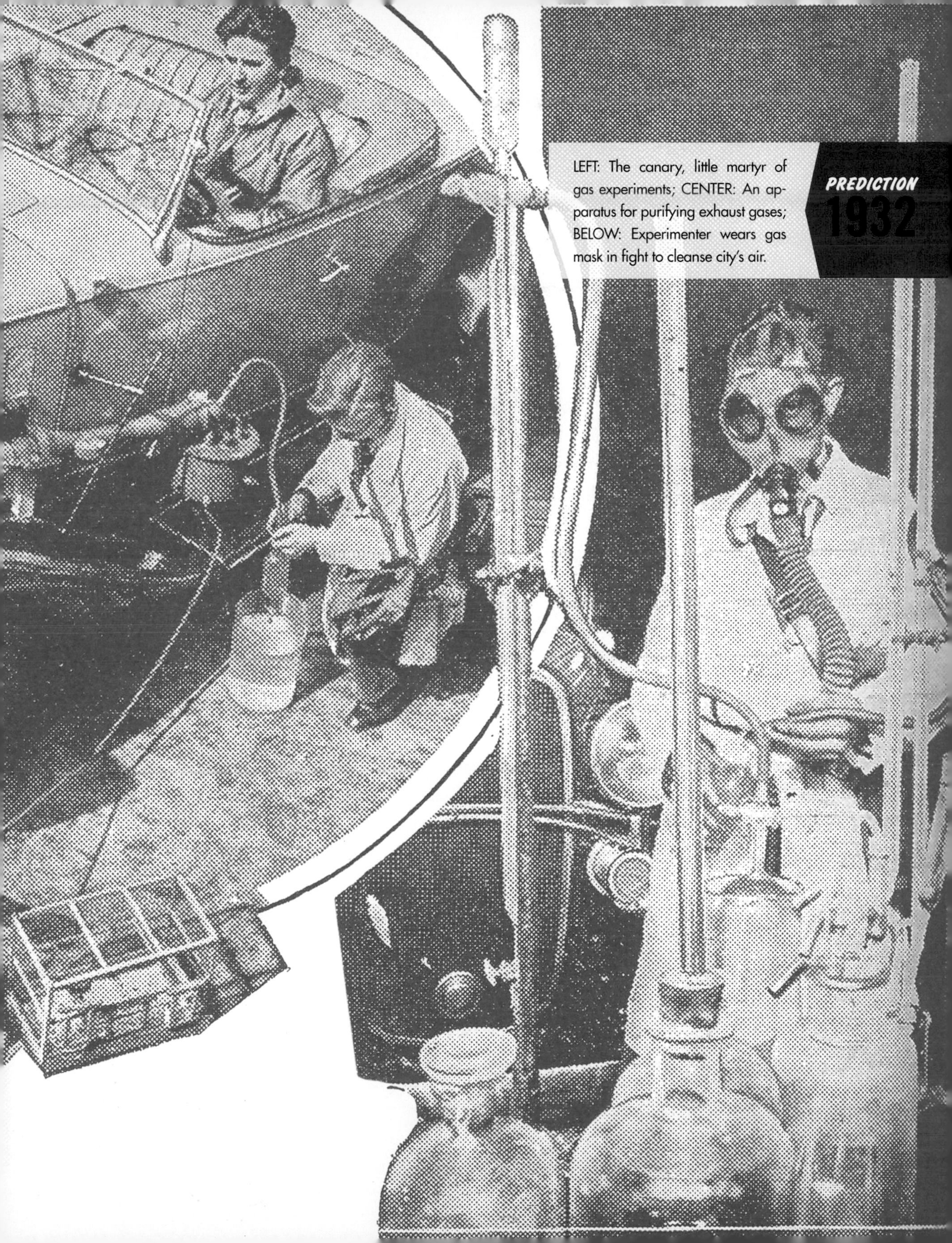

LEFT: The canary, little martyr of gas experiments; CENTER: An apparatus for purifying exhaust gases; BELOW: Experimenter wears gas mask in fight to cleanse city's air.

THE SOOTLESS GARDEN CITY

PREDICTION 1931 Within ten years the modern city will be a comfortable place in which to live, as noise is concerned. This is the prediction of Edward F. Brown, director of the noise-abatement commission of New York City, who believes that in another decade the large metropolis will enjoy silent riveting machines, more melodious auto horns, quieter subways, and sound ventilators for windows which will keep out all unpleasant noises. Buildings of the future, he believes, will be erected to be soundproof just as they are now built to be fireproof.

PREDICTION 1932 In the odorless city of the future, automobiles will wear gas masks. Poisonous and disagreeable fumes from gasoline and oil-propelled vehicles will disappear, making it unnecessary for a city's inhabitants to wear gas masks. By running an engine in a tightly closed room, experimenters protected by gas masks seek to determine how much poisonous matter must be removed from an automobile's exhaust to make it harmless and inoffensive.

PREDICTION 1944 Underground pneumatic tubes to suck household wastes and ashes away may supplant garbage trucks in the city of the future, according to Morris M. Cohn, a sanitary engineer for Schenectady, NY. This "subway" system will eliminate the storage and handling of wastes and would consist of a network of pneumatic ducts under the streets, with a connection in every home, store, and business. An airlock chamber would permit the property owner to discharge wastes into the city's refuse veins, from whence they will travel to an incinerator.

COUNTRYSIDE SOON TO BE INDISTINGUISHABLE FROM CITY LIFE

PREDICTION 1931 "Suburbs of the future will be from fifty to 100 miles from the present centers of the modern cities," said Col. Halsey Dunwoody, vice president of the American Airways system. "Men will live where they want to live and work where they want to work. Just as the whole economic system was changed by our 700,000-mile system of hard-surfaced roads, so the airplane will bring further and even more remarkable changes. The airplane is rapidly removing the letter T from the word there."

PREDICTION 1932 Wireless telephones and televisions will enable their owners to connect to any room similarly equipped to hear and take part in the conversation as

Models of Democra-city, a planned city of tomorrow, as seen by World's Fair visitors.

PREDICTION 1939

easily as if he put his head in through its window. The congregation of men in cities will become superfluous. There would be no more reason to live in the same city with one's neighbor.

PREDICTION 1950 In Tottenville, a hypothetical metropolitan suburb of 100,000, there are parks and playgrounds and green open spaces not only around detached houses but also around apartment houses. At the heart of the town is the airport.

Tottenville will be as clean as a whistle and quiet, as it is a crime to burn raw coal that pollutes the air with smoke and soot. Electric "suns" suspended from arms on steel towers 200 feet high will illuminate this city of the future. There are also lamps which are just as bright and varicolored as those that now dazzle us on every Main Street. But the process of generating the light is more like that which occurs in the sun. Atoms are bombarded by electrons and other minute particles, electrically excited in this way and made to glow.

With regular aerial transportation readily available, the population will become widely distributed. People find it more satisfactory to live in a suburb like Tottenville, if suburb it can be called, than in a metropolis like New York, Chicago, or Los Angeles. Cities will grow into regions, and it is sometimes hard to tell where one city ends and another begins.

You'll eat food from sawdust . . .
shop by picture-phone . .

PREDICTION 1950 The best way of visualizing the new world of A.D. 2000 is to introduce you to the Dobsons, who live in Tottenville, a hypothetical metropolitan suburb of 100,000.

The highways that radiate from Tottenville are much like those of today, except that they are broader with hardly any curves. In some of the older cities, where it was difficult to alter the streets because of the immense investment in real estate and buildings, the highways are double-decked. The upper deck is for fast non-stop traffic; the lower deck is much like our avenues, with brightly illuminated shops. Beneath the lower deck is the level reserved entirely for business vehicles.

In the homes, electricity is used to warm walls and to cook. Factories all burn gas, which originates in sealed mines. The tars are removed and sold to the chemical industry for their values, and the gas thus laundered is piped to a thousand communities. But that's not the only source of energy in Tottenville. Theoretically, 5000 horsepower in terms of solar heat fall on an acre of the earth's surface every day. Many farmhouses in the future will be heated by solar rays and some cooking will be done by solar heat.

2
HOME, SWEET HOME OF
TOMORROW

PREDICTION

1922

The walls between two rooms may be rolled up, like awnings, to throw them together as one.

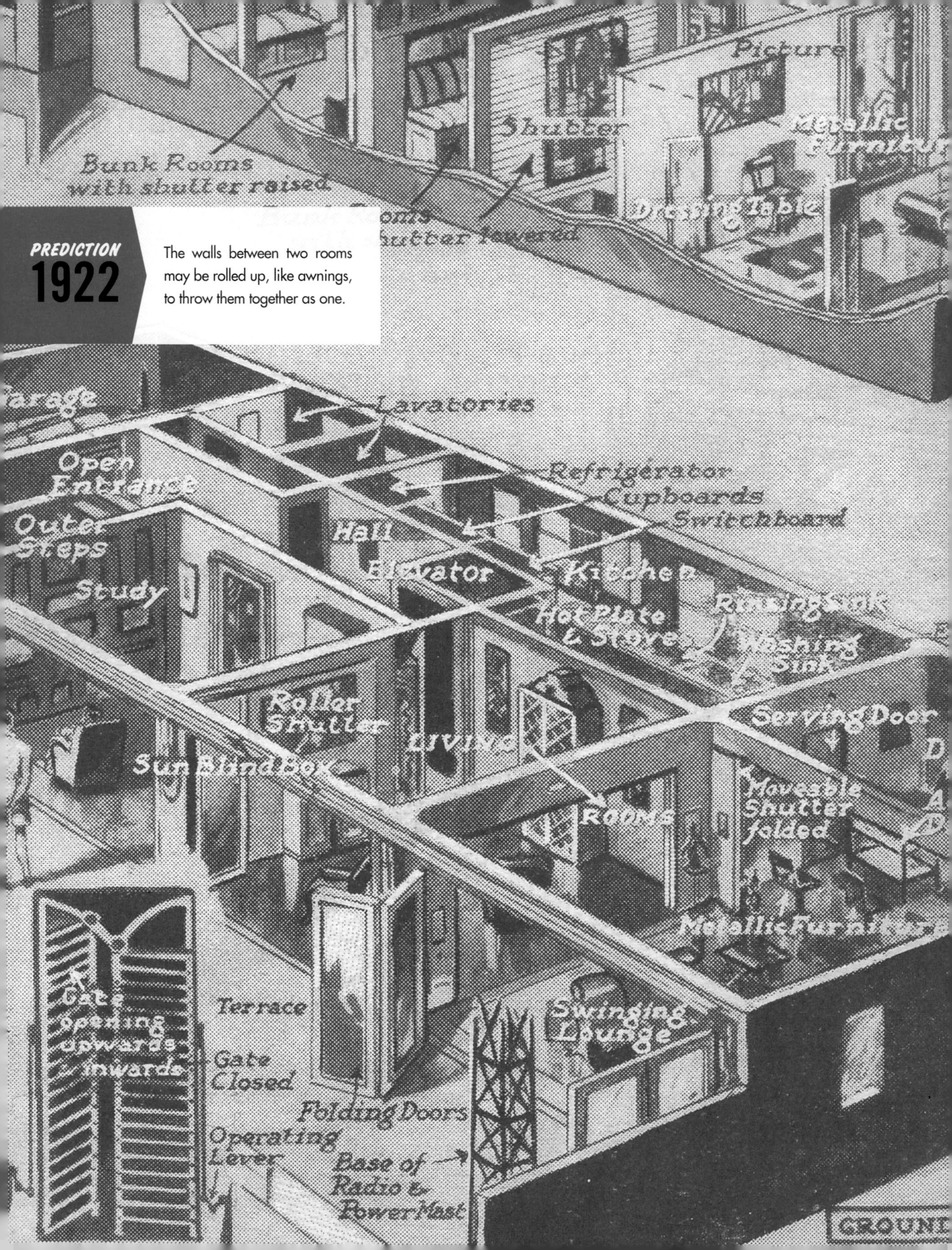

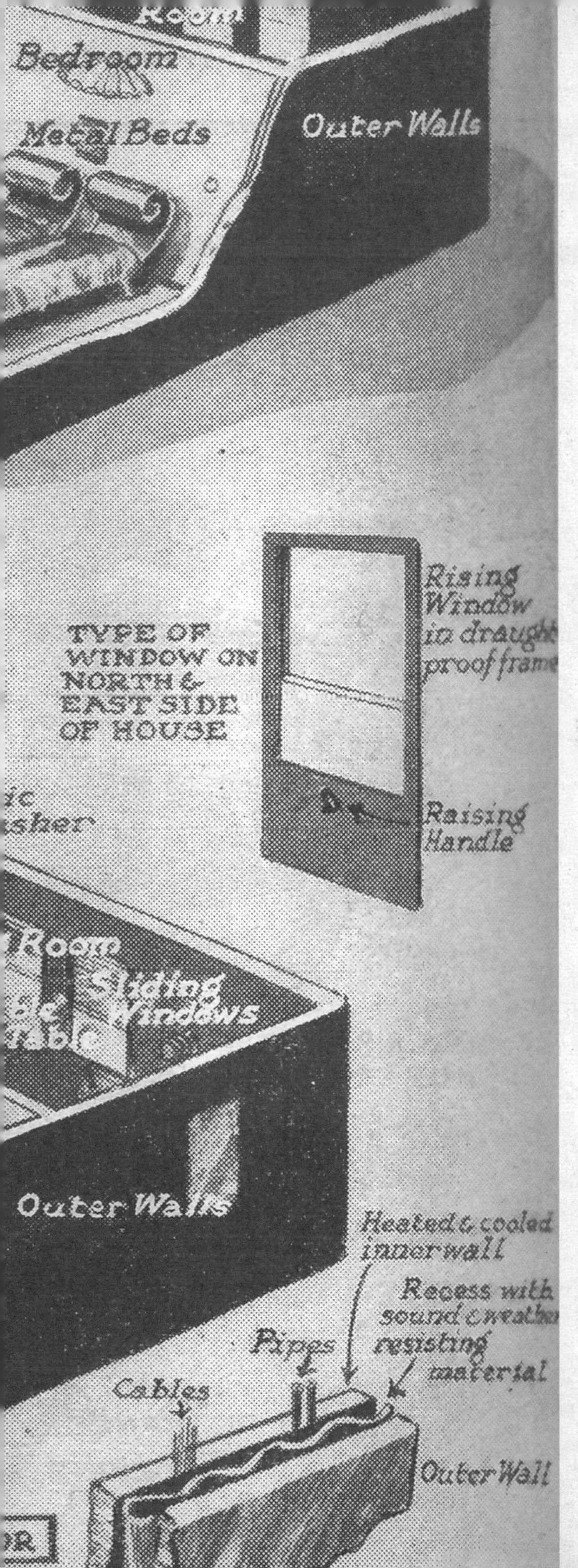

We experience lives of relative luxury now, largely forgetting that only a century ago, housework consumed the lives of women and confined them close to their homes. Men did many more chores than they do now—plumbing, fumigating, chopping wood, hand-churning butter. Ordinary modern conveniences liberated housewives and created new demand for electrical power.

Electricity was *the* transformative technology of the twentieth century. The big emblems of progress—radio, airplanes, cars, TV—all used it, but arguably the greatest impact of the good ol' wall plug was the modern home. Elcctricity, and then electronics, made running a home far easier—often, with music playing.

In the early part of the twentieth century high schools began teaching "home economics" because using the new, easier appliances rushing to market demanded some prior education. So did handling the billing for utilities that spread from city to farm.

My parents had ice delivered for their icebox. The automatic washing machine, dryer, and vacuum cleaner were big improvements over the washboard, clothes line, and broom I saw aplenty in the 1940s. (Ironically, today the avant-garde "homestead" by getting off the electrical grid, once the sign of modernity.) In 1900 there were still more outhouses in America than indoor flush toilets. Now advanced models incinerate waste, rather than using water to flush it away.

The twentieth century future was above all *sanitary*. Flush toilets, antibiotic hand towels (instead of just plain soap, which is just as effective), and many more gadgets sold madly. Wall-to-wall carpeting and sealed-up, air-conditioned buildings didn't just keep out contaminants—they added some of their own, trapping bacteria and giving some sick building syndrome. As it turned out, simple, clean water was the biggest benefactor of better health, but it's hard to wax rhapsodic about it.

The "vidphone" of the 1920s and 1930s didn't make it in the marketplace, though some were sold. Instead, the Internet gave us all free video-phone access, via Skype. We know enough now not to show off bed head or sloppy clothes. This is a good example of a frequent phenomena—predictors got the technology essentially right,

PREDICTION 1956

EXPERIMENTAL *"PICTURE PHONE"*

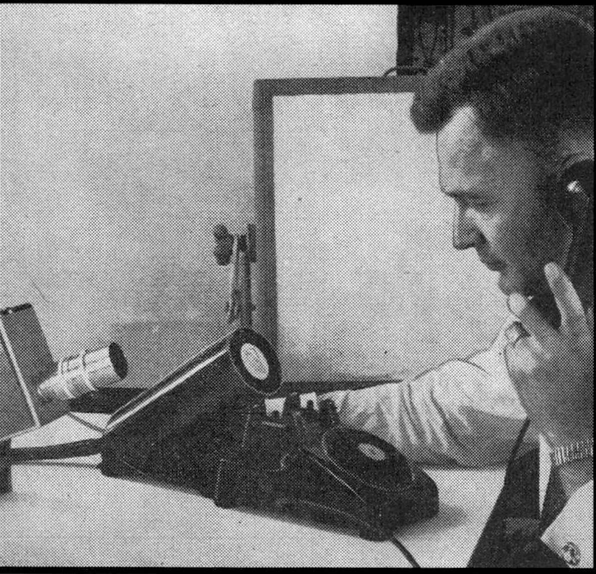

Soon you may be able to see the person you are talking to on the telephone. Bell Laboratories has developed an experimental "Picture-Phone" system which has transmitted pictures over short and long distances, even from New York to Los Angeles. The phone system sends out only one picture every two seconds, and the picture has less detail than television. It shows promise of being commercially feasible because the picture can be transmitted over a pair of ordinary telephone wires instead of expensive coaxial cables or microwave relays. Instead, one new line, consisting of a pair of wires like the regular telephone line, would be installed at your home to carry the picture.

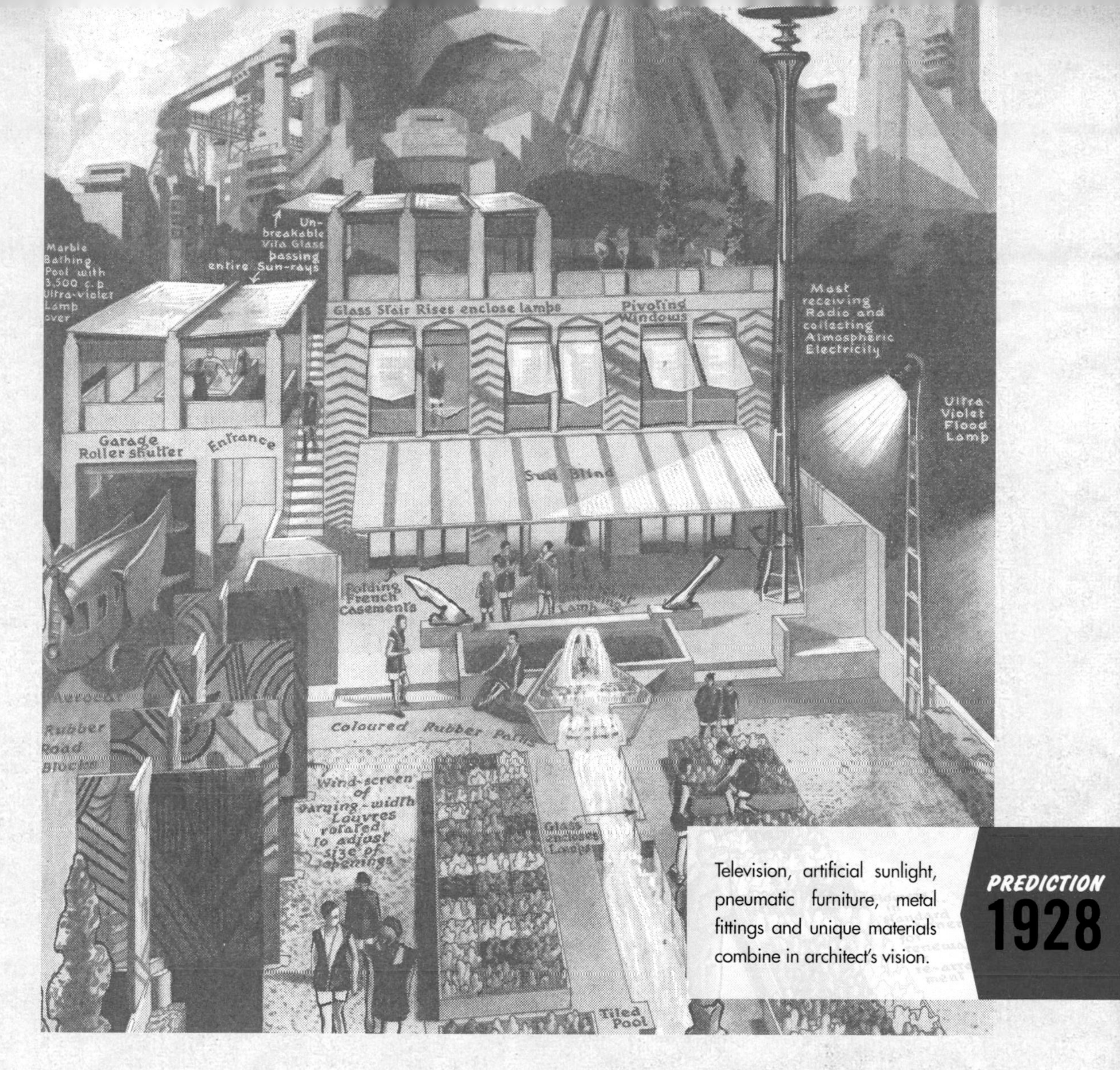

PREDICTION 1928

Television, artificial sunlight, pneumatic furniture, metal fittings and unique materials combine in architect's vision.

but not the marketing. Nobody who wrote of vidphones in science fiction or *Popular Mechanics* thought they would be free.

Answering machines, carbon paper, photocopiers, electric typewriters, and finally the computer made information easier to share, whether at home or in the office. But the computer outdid them all, and now it can control homes directly, managing temperature, lights, and appliances. The "servant problem" of the early twentieth century got solved by smart, tireless software and home robots like the Robovac.

Surprisingly, women didn't decrease the hours they spent on homemaking. Why? Because people opted for *quality*—more or better meals, cleaner clothes, more attractive gardens. They wanted a future that was not just bigger, but better. And they got it. ◎

PREDICTION 1922 "Sounds like a good buy!" says the customer of the Walter Teague model in her hands. It would sell for $2,000 or less and—unlike Rome—this prefabricated unit could be built in a day.

A LAND OF PERPETUAL SUNSHINE

PREDICTION 1928 The home of the future, as an architect envisages it in year 2000 was exhibited in London recently. The new invention Vitaglass admits the sun's ultra-violet rays in fair weather into each room, and produces artificial sunlight for cloudy days and night use to create a permanent summer-day effect.

Complete with convertible metal furniture; bunk rooms rather than bedrooms, laid out somewhat like steamer cabins; movable walls; a garage for a combination airplane-automobile with folding wings; and, on the garage roof, a second-story swimming pool, gardens fitted with plants in movable containers; and wireless power and program reception masts, the citizens of tomorrow will have a home full of sunlight.

PREDICTION 1942 Look ahead to your postwar dwelling five or ten years to picture possibilities like these:

Some Sunday morning you decide you would like to live on a piece of land which has caught your fancy. It is a bare plot in a developed real estate area with sewage, water, electricity and gas connections. Monday morning you close the deal. Preconstructed foundation piers may be set into the bare plot and utility pipes extended to the proper points. By Monday noon you have signed up for the house for Tuesday delivery.

PREDICTION 1935

HITTING THE TRAIL

From the prairie schooner to the modern trailer coach is a vast step forward. More surprising is this prediction of Roger W. Babson, able statistician: "Within twenty years, more than half the population of the United States will be living in automobile trailers!"

A HOUSE *BLOOMS LIKE A FLOWER*

Buckminster Fuller, architect and inventor, has constructed a house built around a mast. Says he: "This house is built from the top down, and laying the foundation is one of the last things to be done. The mast is placed at the exact center of your house-to-be... The entire operation from placing the mast to moving in your furniture takes only a few hours."

Tuesday morning a truck or two pulls up to the site with half a dozen men. While you are downtown selecting whatever new furniture you may need, for Wednesday afternoon delivery, the small construction crew sets up your new home.

Wednesday, surely by noon, you can move in your old furniture before the truck comes with the new. If you have found time to shop for food, Wednesday's evening meal can be eaten in the new abode.

This is not all. Perhaps only a few months later a shift occurs in the family setup. A son or daughter leaves home to be married, and a member of the family decides to take up photography. You advertise in a newspaper: "Wanted to exchange on bedroom, style X2-A, of the Jones Home Corporation line, for a darkroom." A deal is made. A truck brings the prefabricated darkroom interior, pulls out the prefabricated bedroom interior and the home has changed to fit your family.

Paul Nelson, famed architect, envisions this prefabricated shell in which prefabricated interiors can be replaced or changed. He says: "The reason we could have such seeming luxuries is that mass production could provide them at a cost almost unbelievably low. The same evolution would surely occur with mass-produced rooms that occurred with the automobile. There would be the possibility of buying and selling second-hand rooms. This would bring these superior living units within the reach of even the lowest income groups.

"The process would be almost as simple as plugging in the connections of your refrigerator or washing machine."

PROOF AGAINST FIRE, VERMIN, AND WEATHER

PREDICTION 1928 **The walls of the house of the future will be made of a hornlike substance,** currently in the experimental stage in a London laboratory. It is tough, impervious to moisture and capable of being cut and welded at high temperatures. The foundations and framework of the house will be of stainless steel. The wall material can be colored or produced in any desired patterns.

PREDICTION 1937 **Houses of the future will be built of plastic and synthetic materials** that should outlast materials used today, industrial chemists predict. From new applications of plastic materials, lacquers and synthetic fibers, new types of homes will emerge, with building costs much lower. Such houses would be noiseless, sanitary and proof against fire, vermin and weather.

PREDICTION 1946 **Unlike most new industries, conservatism reigns among the so-called "brass hats" of plastics.** Rash statements about all-plastic houses and

PREDICTION
1928

An inches-deep rooftop lake may become an important air-conditioning method.

automobiles are frowned upon. Leaders in the industry see the future of plastics as that of a new material ranking in importance with many metals, woods and glass, but not necessarily replacing any of them.

PREDICTION 1950 By A.D. 2000, metallurgical research will heavily influence both civil engineering and architecture. Steel will only be used only for cutting tools and for massive machinery. Ways will be found to change the granular structure so that a metal is ultrastrong in a desired direction and weaker in other directions. As a result, the framework of an industrial or office building or apartment house is an almost lace-like lattice.

Thanks to these alloys, to plastics and to other artificial materials, these houses differ from those of our own time. A typical house has light-metal walls only four inches thick. There is a sheet of insulating material an inch or two thick with a casing of sheet metal on both sides. By 2000, wood, brick and stone are ruled out because they are too expensive.

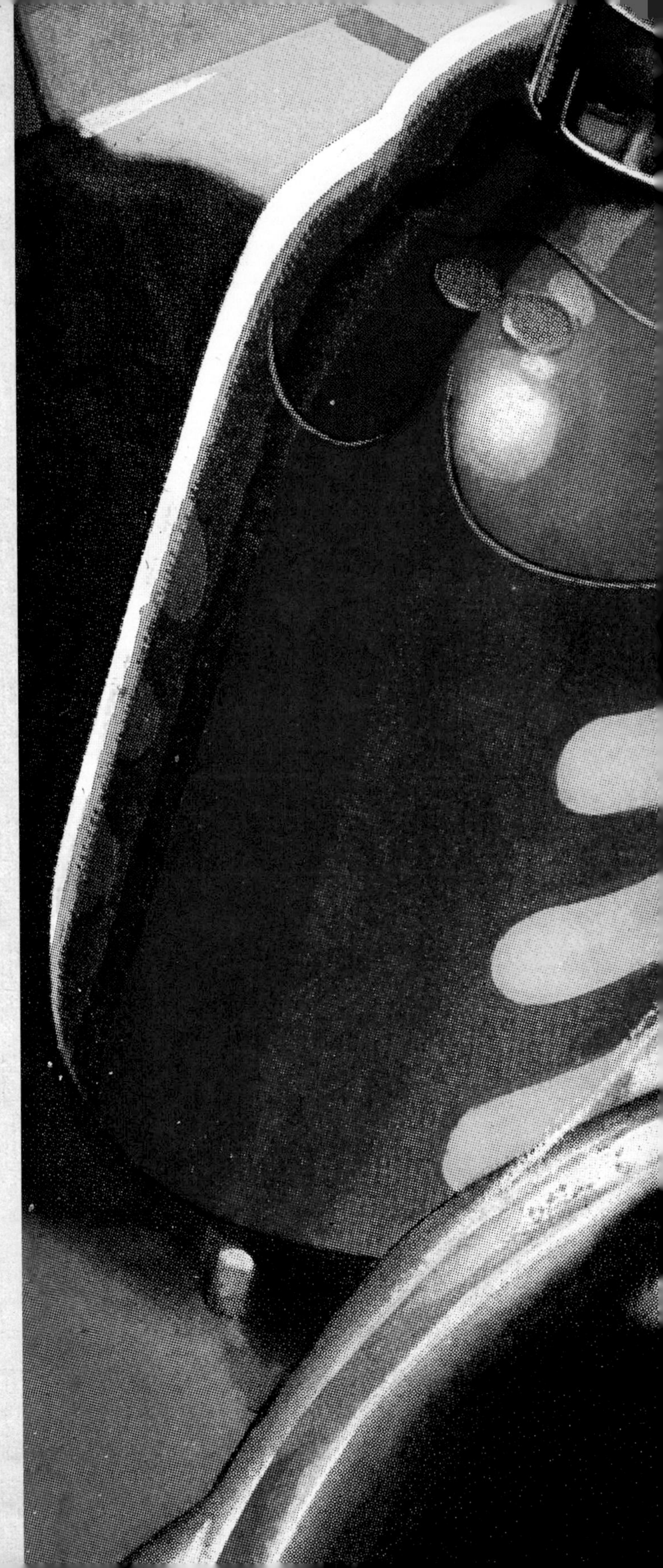

Safety first! The bikes of the future will be encased by a strong plastic or plastic-like substance to protect racers and recreational cyclists from harm.

PREDICTION
1933

THE LAST WORD IN FUTURE FASHION

PREDICTION 1913 Dr. William H. Perkin, the English chemist who has done much remarkable work on the synthesis of rubber, has now found a method of making cotton cloth absolutely fireproof. At the same time it is softer and more beautiful than the cotton cloth from which it is made, and, the inventor believes, free from the objections to fireproof cloth heretofore manufactured. Doctor Perkin showed one garment made of it which was unharmed by 20 washings, and another which had been worn for two years and washed every week. Neither could be burned, though the latter was worn almost to rags.

Not only will this fireproofed material be found suitable for the manufacture of clothes for firemen, but it may also be used by women for the finest party dresses, the most delicately colored fabrics being unharmed by the process and even given a softer appearance. For children's clothes it should be particularly desirable, since it islikely to prevent many accidents.

PREDICTION 1913 Future clothing and accessories will be made entirely out of plastic. This Saran purse wipes clean with a damp cloth. Rainbow-colored plastic buttons put life into any outfit.

PREDICTION 1913 That an august body of scientists such as was gathered at the recent International Congress of Applied Chemistry should concern itself with future fashions for women's dress may seem peculiar, but one of the most celebrated lecturers, Giacomo Ciamician, made pre-

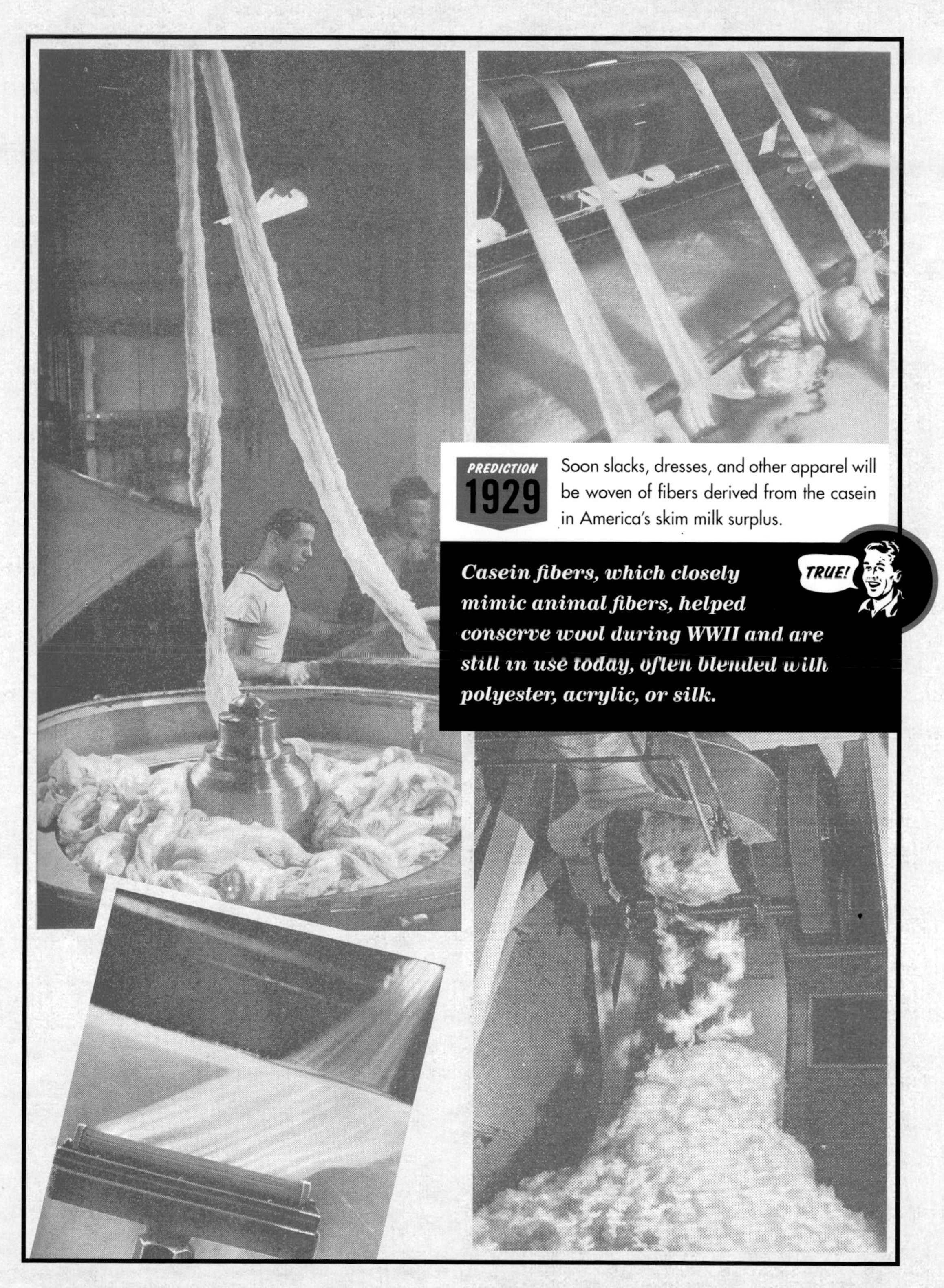

PREDICTION 1929

Soon slacks, dresses, and other apparel will be woven of fibers derived from the casein in America's skim milk surplus.

TRUE!

Casein fibers, which closely mimic animal fibers, helped conserve wool during WWII and are still in use today, often blended with polyester, acrylic, or silk.

Colorful weaves give plastic fabrics a pleasing appearance, and they have a washable finish.

PREDICTION
1946

dictions as to future fashions. Women of the future, he said, will no longer be contented with a dress which remains constantly of one color, but will demand colors that change in harmony with their surroundings. Thus the color of the apparel may be changed without changing the dress. Passing from darkness to light the color would brighten up, conforming automatically to its environment.

This prediction will come true as soon as chemists learn to understand better what are called "phototropic colors," or colors that change with the intensity of the light upon them. In men's wear this might mean that the light-colored suit of the bright summer day could be transformed into a dark suit at night.

PREDICTION 1929 Dresses of asbestos that will be as lustrous as silk and will give long wear, with ease in cleaning, are predicted by an eastern scientist. Fabrics are already being made from trees and vegetables and the Romans made a sort of cloth from asbestos fibers centuries ago, so this prophecy is considered entirely reasonable by experts. The use of asbestos in the early Roman days was confined largely to the weaving of shrouds. According to tradition, Charlemagne had a tablecloth of asbestos which was cleaned by throwing it into the fire. In the seventeenth century, Chinese merchants displayed asbestos handkerchiefs and the Eskimos in Labrador have used lampwicks made of an asbestos fabric for many years.

PREDICTION 1929 Clothing made of aluminum will be in vogue within the near future, a German metallurgist predicts. He points out that the material is already being used successfully to cover shoes, pocketbooks and vanity bags and is also being employed in the interior decoration of airplanes. The metal, in the form of thread, can be stretched through layers of cellulose superimposed upon it and subjected to pressure, and is not likely to tear or crumple, the expert declares. Considerable quantities of aluminum brocades, for women's garments, are being exported.

PREDICTION 1950 Despite those gloomy predictions that our oil reserves will soon be exhausted, Dr. Gustav Egloff of the Universal Oil Products Company believes that in the year 2000 we will be refining 16,000,000 barrels of petroleum daily. He predicts that most of our clothing in the next half century will be made basically from petroleum.

THE HOUSEKEEPER OF THE FUTURE

PREDICTION 1938 Dust-free air, purged of 99 percent of the microscopic dirt and bacteria floating in it, is promised for the not distant future by a "dust magnet" developed by Westinghouse Electric and Manufacturing company engineers. In the dustless home of tomorrow it is anticipated that annual cleaning of wallpaper and frequent washing of curtains and rugs will be eliminated. In a test of the electrostatic system at Pittsburgh, curtains remained clean for months. A particle of cigarette smoke is about four millionths of an inch in diameter. Billions of them are blown into the air with every puff. But they are not too small or too numerous for the electric dust magnet to tackle!

PREDICTION 1942 Sterilizing air by means of ultraviolet light produced artificially is destined for a rapid increase predicted Dr. Theodore S. Wilder of Philadelphia, Pa. Ultraviolet light is fatal to bacteria and apparently to the viruses which are so small they cannot be seen under a microscope. Private homes can apply the rays to sterilize air in the nursery, and ultraviolet light might also take the place of the sheet, soaked in antiseptic, which used to be hung across the door of a sickroom. However, Dr. Wilder warned that the need for ultraviolet light in each case should be determined by a physician and its installation handled by experts.

PREDICTION 1950 When the housewife of 2000 cleans house she simply turns the hose on everything. Why not? Furniture (upholstery included), rugs, draperies, unscratchable

PREDICTION 1910

IRONING CRACKS FROM A DAMAGED ARMCHAIR

It may be possible for the housekeeper of the future to iron out the scratches and dents in her furniture as she does the creases and wrinkles in her clothing. The process simply requires the filling of the crack or nick with paste, after which an ordinary electric flatiron is passed over the spot thus treated. This causes the wood to swell, and the rubbing or smoothing with the iron restores the original surface. The result is that the damaged piece of furniture regains its original finish.

floors—all are made of synthetic fabric or waterproof plastic. After the water has run down a drain in the middle of the floor (later concealed by a rug of synthetic fiber) she turns on a blast of hot air and dries everything. A detergent in the water dissolves any resistant dirt. Tablecloths and napkins are made of woven paper yarn so fine that the untutored eye mistakes it for linen. She throws soiled "linen" into the incinerator. Bed sheets are of more substantial stuff, but she has only to hang them up and wash them down with a hose when she puts the bedroom in order.

PREDICTION 1957 In A.D. 2000, our comfort environment will be so well controlled that we will be able to keep the atmosphere at the ideal level for the happiest, most energetic, productive life. Houses will be kept so clean by electronic dust and dirt traps that housecleaning will never be necessary. Dining-room tables will quietly swallow dishes after a meal and transfer them to a dishwasher which will clean the dishes, dispose of garbage, stack and store eating utensils until the next mealtime.

PREDICTION
1950
Because everything in her home is waterproof, the housewife of 2000 can do her daily cleaning with a hose.

EVOLUTIONARY CHANGES IN HOME EQUIPMENT

PREDICTION 1918 During a recent convention of bakers in London, an apparatus was demonstrated which utilizes superheated air for cooking purposes. The steam of an ordinary kitchen boiler is conveyed in pipes to the super-heater, where in a series of coils above a coke fire, its temperature is raised to 1,000 degrees Fahrenheit without increasing the pressure. Passed through the hollow rods of a griller, the air quickly raises the metal to the same temperature as itself, and then anything can be cooked from a steak to biscuits.

PREDICTION 1963 A glass-dome oven, a range that cooks by induction heating without warming its marble top and a refrigerator with revolving shelves are features of this kitchen. The uncooled top section of the refrigerator stores dry foods.

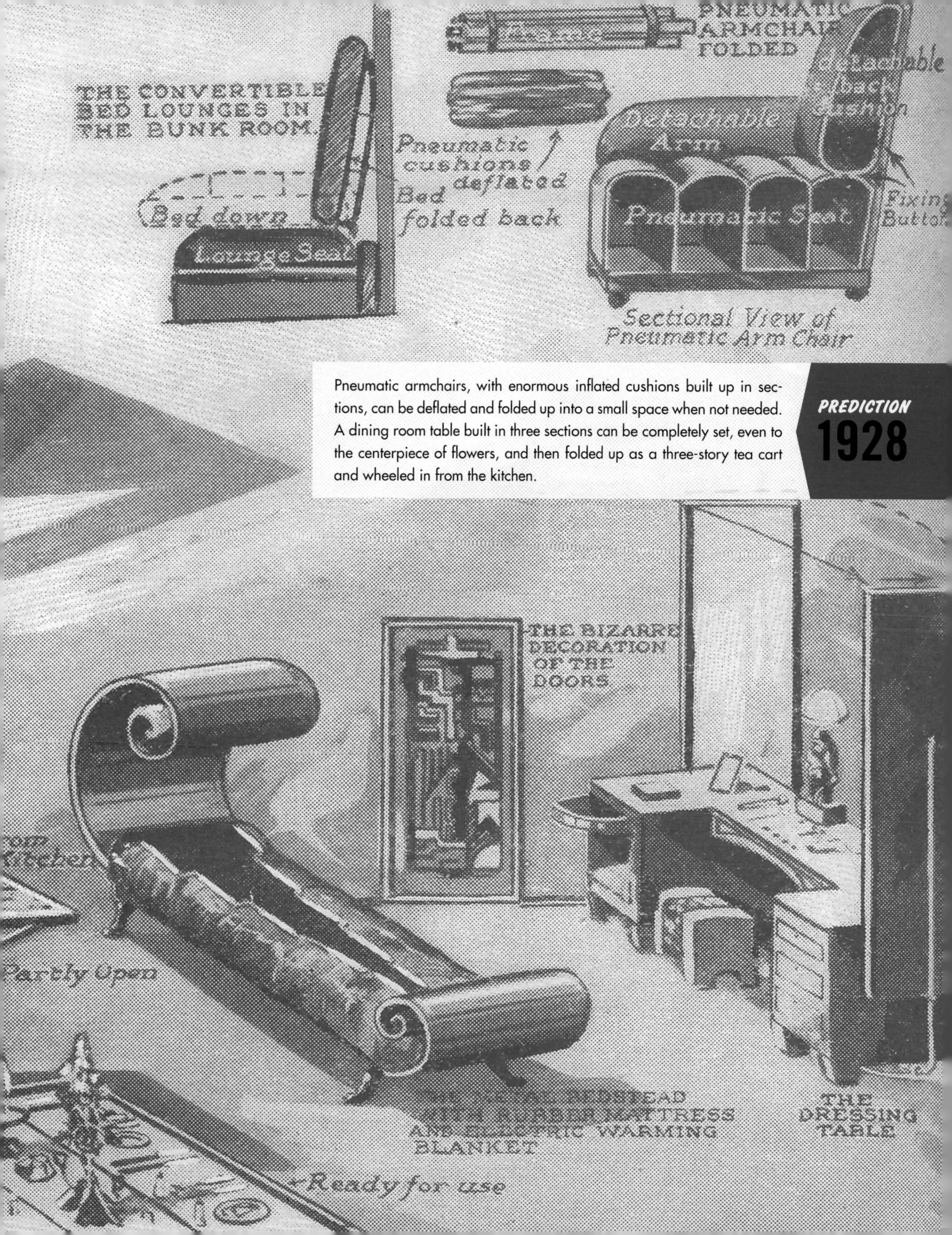

Pneumatic armchairs, with enormous inflated cushions built up in sections, can be deflated and folded up into a small space when not needed. A dining room table built in three sections can be completely set, even to the centerpiece of flowers, and then folded up as a three-story tea cart and wheeled in from the kitchen.

PREDICTION 1928

PREDICTION

1948

SILENT SOUND SHOWS ITS MUSCLE

Sound too high pitched to be heard may become the washer-woman of the future. Experiments at Pennsylvania State College have proved that ultrasonic waves wash dirty clothes better than present laundry methods.

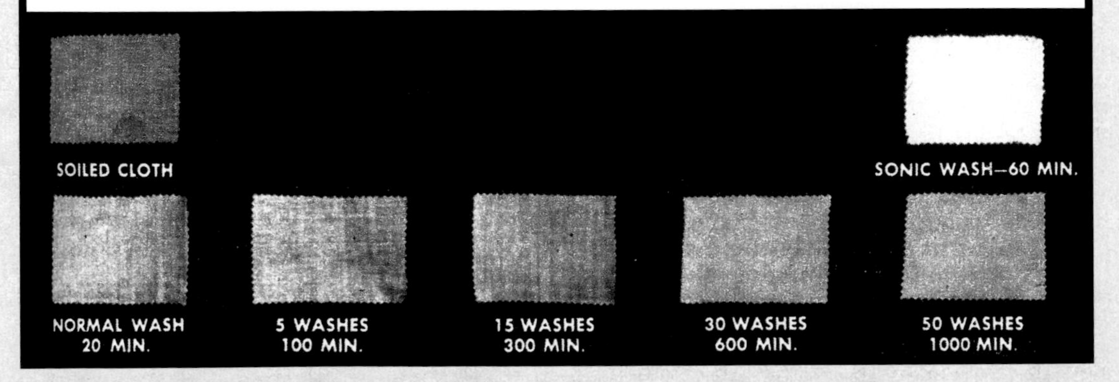

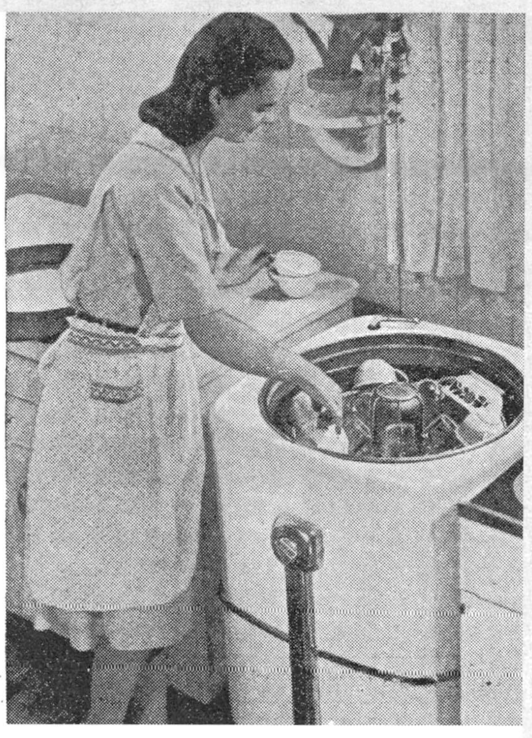

PREDICTION 1945 This housewife uses a clothing washer with a basket arrangement that holds service for six persons.

The kitchen of the future will have no blazing fire, and that everything will be cooked by the turn of a valve admitting the heated air. The air, after passing through the hollow grillers, can be further utilized for heating rooms.

PREDICTION 1944 Westinghouse looks forward to completely air-conditioned homes with dust and dirt collected from the air electrostatically by home-size Precipitrons; lockers to quick-freeze foods from your post-victory garden; automatic washing machines with electric driers and ironers; and more.

Furniture and textiles are bound to undergo changes as molded plywoods and plastics and synthetic fibers have their first chance at peacetime competition. Sundberg and Ferar have suggested a plastic chair designed for comfort, its body molded by the "heatronic" method with large presses. Instead of clumsy steel springs, the designers propose a preformed sponge-rubber cushion and a zipper fastening the upholstery of woven extruded plastic to the chair. The back and legs would be transparent, and the chair would be shaped to support the body at every point.

PREDICTION 1926 This housewife finds frozen foods cook more quickly.

PREDICTION 1928 **Fifty years hence, according to Roger W. Babson, internationally known statistician, the milk bottle will probably be a museum relic,** along with the ice wagon, the coal shovel and the ash can, and our milk and butter will be derived from kerosene instead of cows, while most of our other food will be served in concentrated or pill form.

PREDICTION 1952 **Dr. Lee de Forest, the "Father of Radio," foresees great refinements in the field of microwave technique,** whereby living rooms and their occupants will be heated by high-frequency waves from

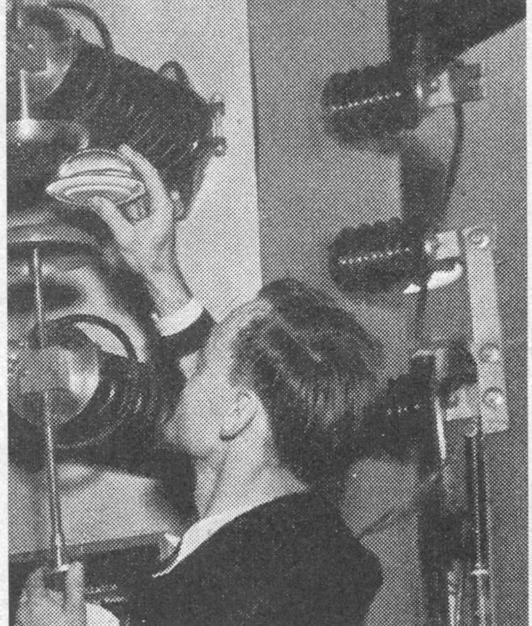

PREDICTION 1937 Cooking a ham sandwich in high-frequency radio waves. This method may be common in the home of the future.

PREDICTION 1969

NOW THAT'S A *BATHROOM!*

Visitors to the "Building 68" exhibition in Munich could understandably have gone away believing the bathroom will become the activity center of the home. The bath of the future has a sunken tub for bathing and swimming. Around the wall (the room is circular) are niches for makeup, television, a small library, a bar and a rest corner.

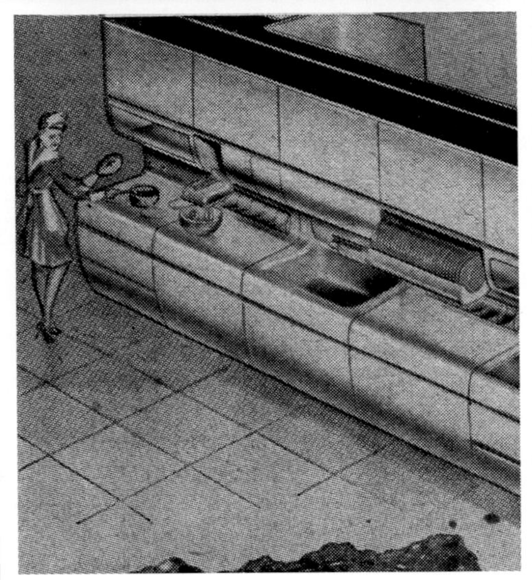

This kitchen from Sundberg & Ferrar has a built-in refrigerator with drawers instead of doors. The sink, recessed into the wall, includes a plastic shower head for rinsing dishes, its hose automatically drawn up on a hidden reel when released. On the other side of the unit is the bathroom, with built-in fixtures conveniently arranged.

walls or ceilings, and short waves will be generally used in the kitchen for roasting and baking, almost instantaneously, with far better results than in the past.

PREDICTION 1956 Like the dream cars of the motor manufacturers, a dream kitchen unveiled by Frigidaire has a host of startling features. Dishes are washed in three minutes by ultrasonic waves. The self-rinsing sink features a warm-air hand drier, power-driven pot scrubber and a control that measures out a given amount of water heated to a desired temperature. A cylindrical refrigerator with rotating shelves can be loaded from inside or outside the house.

A hands-free, distant-talking TV telephone features a device that automatically dials any of 50 most frequently called numbers when a button is pressed. It can be set so that by dialing home, you can start the oven, open or close windows, or perform other operations. A TV tube would show the faces of callers, views of the nursery or front door, or regular TV programs. A motorized serving cart can be moved by remote control.

PREDICTION 1963 This housewife's dream was designed and built in his spare time by a suburban engineer in Minneapolis. His wife may simply flick a switch on the end of the "diner" counter and two table leaves fold down with dishes and silverware already in place for a meal.

When mealtime is over, an electric switching mechanism is dialed to activate a dishwasher, a garbage disposal button is pressed and—presto!—all the drudgery is removed from kitchen cleanup. Even the dining table is all set for the next meal.

DINNERS WITHOUT *DRUDGERY*

Before long you may see frozen dinners served in hotels, trains, planes, ships, factories, offices and your own home. They probably will be sold in grocery stores and delicatessens.

A wide selection of frozen dinners is expected to be available soon in grocery and frozen food stores. Eventually, frozen meals will be delivered from house to house.

TRUE!

The first TV Dinner was sold by C.A. Swanson & Sons in 1953, spawning a huge frozen-meal industry.

YOU'LL EAT GRASS AND LIKE IT

PREDICTION 1926 **Research will provide a synthetic food from coal** when needed and abundant synthetic fuels for motors long before the natural oil deposits are exhausted, in the opinion of Prof. James F. Norris, president of the American Chemical society. The Germans prepared food from coal during the war, he points out, and fats were made from petroleum. While these accomplishments are regarded as chiefly of interest to the scientist, they suggest a practical solution of food and fuel shortages, should these emergencies arise. Prof. Norris declared that food supply will never become an acute problem, so long as we have chemists.

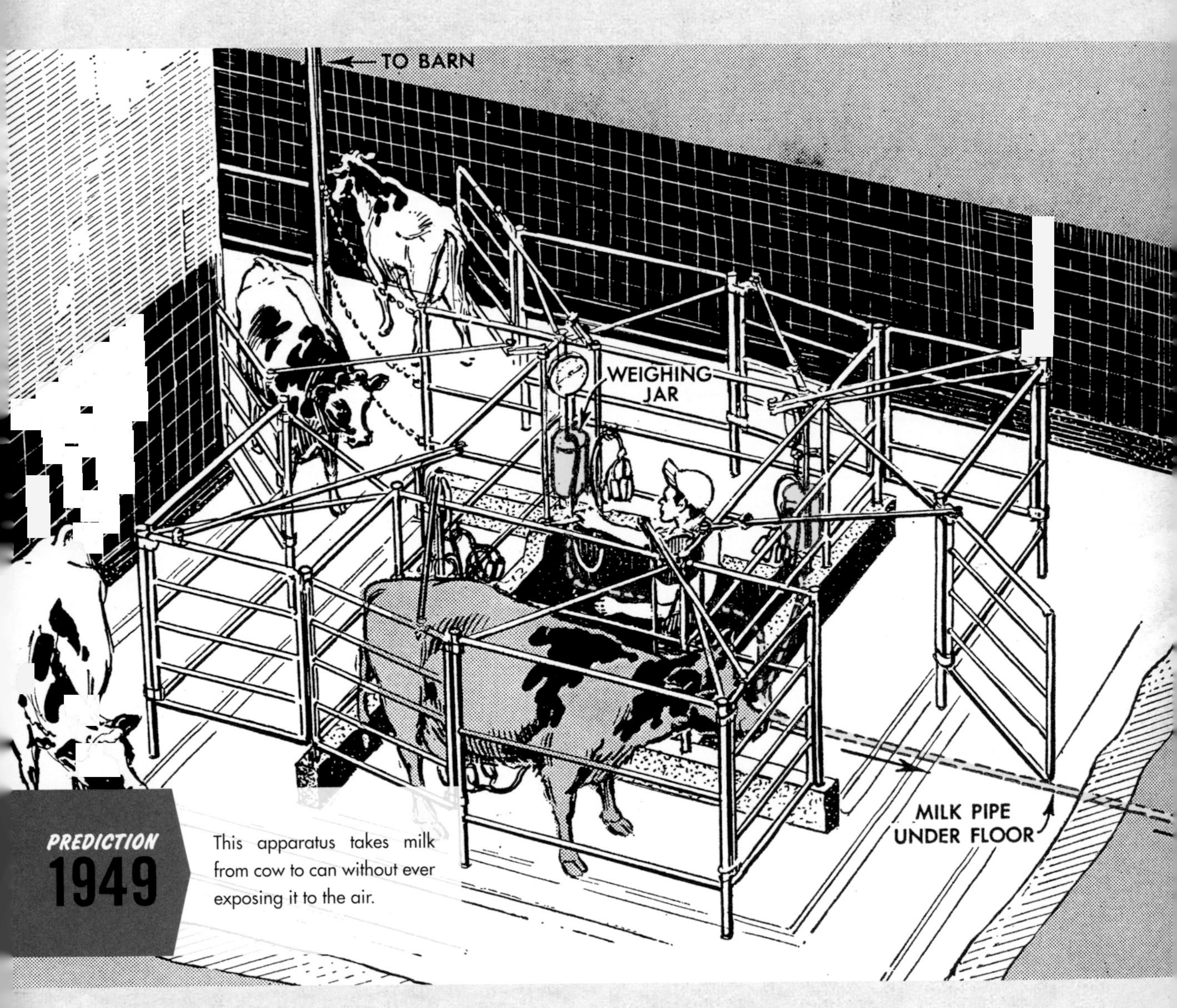

PREDICTION 1949 This apparatus takes milk from cow to can without ever exposing it to the air.

PREDICTION 1937 **The grocer's store of the future will deal largely in foods kept fresh by freezing.** Most of the shelves and storage space will be refrigerated, all electrically operated.

Shoppers will buy and take foodstuffs home to conveniently-sized cold lockers. The range of foods will include all types of meats, fish, poultry, vegetables, fruits, salads and juices—from frozen spinach and frozen eggs to frozen mushrooms and frozen steaks. Frozen foods should touch new highs in sanitation, containing no micro-organisms more harmful than molds—and not much of that. Whether harvested the day before or fifty years before, every item should be absolutely flavor-fresh and true of color. Superfine fruits, vegetables, and meats, grown by experts or crop specialists, might continue to delight palates generations or even a century after the grower's death.

When the housewife buys a pound of foodstuff she will have a pound of edibles, free of wastes in fat, bone, hulls, stems or pits. Work of preparing food will be cut to a small fraction of what it is today. More foods will be eaten raw and, in the case of vegetables and fruits, cooking time will be reduced from a half to two-thirds. Butchering, drawing and other preparations which now require expert labor in the store will be done at central packing points. Great savings will result from the fact that freezing equipment is compact and portable. This will do away with the present majority of food transportation costs and will erase the necessity of building expensive canning or packing plants which can be operated only a few weeks or months of the year.

PREDICTION 1940 **On the menus of tomorrow, you're likely to find breakfast foods, pancakes, bread, and cookies containing quantities of grass;** there'll be grass in butter, milk shakes, and ice cream, perhaps even in candy bars.

This forward step in nutrition is envisioned by a group of biochemists and scientists whose researches show the vast vitamin content in young cereal grasses can be processed into food without changing its appearance of palatability. Medical science is watching this development with interest.

In twelve pounds of dried grass leaves, Doctor Graham points out, can be found all the vitamins and minerals the body needs, when added to low-cost diets. In this way, a minimum adequate diet can be converted into a liberal one at no appreciable rise in cost.

When the chemists first began working with the idea of putting grass into everyday foods they ran into the color problem. Bread, for instance, with powdered green grass added to the ingredients took on an unappetizing hue. However, by removing the chlorophyll, the green coloring matter in plant life, the powder assumed a

PREDICTION
1940

Ordinary grass gives promise of providing low-income families diets more abundant in vitamins than are now enjoyed by the wealthy. Housewives soon may add nourishing powdered grass to recipes.

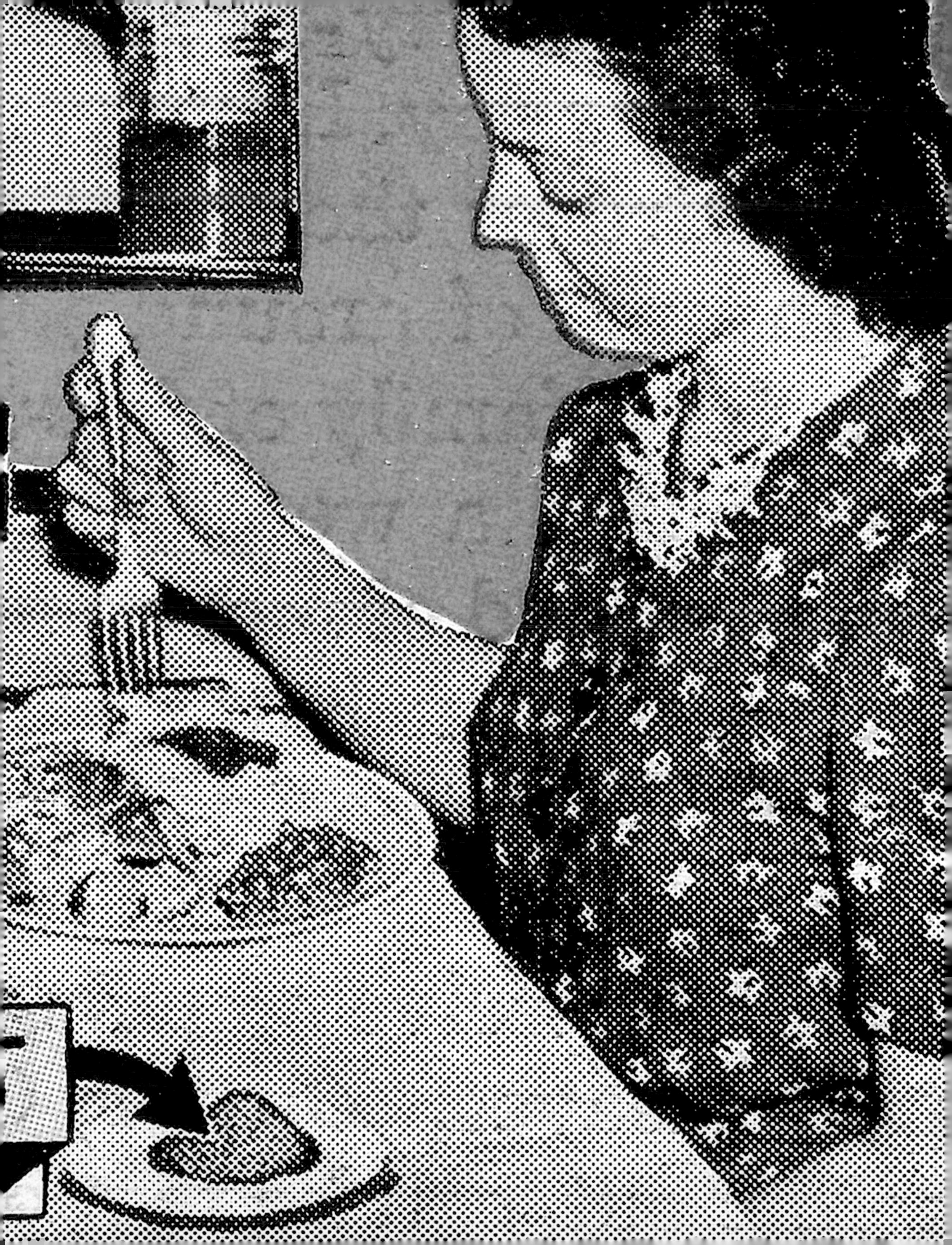

neutral yellow color that looks and tastes like malted milk, and what's more, has an even greater vitamin potency than is otherwise the case. It gives white bread a golden color.

PREDICTION 1950 In A.D. 2000, cooking as an art is only a memory in the minds of old people. A few die-hards still broil a chicken or roast a leg of lamb, but the experts have developed ways of deep-freezing partially baked cuts of meat. Even soup and milk are delivered in the form of frozen bricks.

Some of the food this housewife buys is what we miscall "synthetic." In the middle of the 20th century statisticians were predicting that the world would starve to death because the population was increasing more rapidly than the food supply. By 2000, a vast amount of research has been conducted to exploit principles that were embryonic in the first quarter of the 20th century. Thus sawdust and wood pulp are converted into sugary foods. Discarded paper table "linen" and rayon underwear are bought by chemical factories to be converted into candy.

PREDICTION 1962 Soon you may be sprinkling peanut butter on a slice of bread or munching a cookie that's made from sawdust flour, which is totally lacking in calories.

Cellulose, an organic compound constituting the structural stuff of wood and grasses, has been put into edible form which can be readily digested by man. In its natural state, cellulose is fibrous and resists man's earnest efforts to digest it. Cellulose is noncaloric and provides no nourishment. Inside the body, it takes up space, but since it is not digested, it cannot add weight to the eater.

Two forms of digestible, nonfibrous cellulose have been developed with in high polymer chemistry by American Viscose scientists. Avicel, a highly purified form of cellulose, is a snow-white, free-flowing edible flour that resembles sawdust. A gel can be prepared by dispersing the powdered Avicel in water. It is smooth, odorless, tasteless, easily spreadable and noncaloric. Because of gel-like properties—stability, body, bulk, opacity, texture and palatability—Avicel is an effective ingredient in the preparation and calorie control of dressings, puddings, custards, spreads, aerosol-whipped toppings and frozen desserts.

AN ELECTRIC HOME

PREDICTION 1967 There have been many predictions that the home of tomorrow will be radically different from the home of today, because it will be run by a computer. But no one knew precisely what a computer could do in the home until Jim and Ruth Sutherland of Pittsburgh, Pa., designed, built and programmed ECHO IV. ECHO means Electronic Computing Home Operator, and building it has been a family affair.

Before the end of the second year what was formerly the family basement playroom had been taken over by the home computer and its peripheral equipment and Ruth was wondering, "Will it replace *me*?" She isn't worried about this now, since she has learned that *home* computer programs must first be flow-charted by someone who knows *homemaking*. The Sutherlands feel that if the homemaker programs some of her own tasks, she will better understand how the computer operates and become skilled in determining best household applications.

Now that the basic routines are running in ECHO IV, the Sutherlands delegate the chores of bookkeeping to the computer. Tabulating monthly budgets and accounting for monthly expenditures is a time-consuming job that ECHO will reduce to a

Foods and other items will be carried to the cashier not by the customer but by a conveyor belt in this "assembly line" grocery store. The customer is given with a roll of tape which is punched with holes when she inserts a key in a slot next to the item selected. When she finishes shopping, she hands the tape to a clerk who operates a combination "translator" and adding machine. This instrument interprets the punched holes just as a piano player plays from a music roll. Electrical impulses race to gravity chutes and release guards that drop unbreakable articles to the conveyor belt. More delicate merchandise is lowered to the moving belt by a tripper shelf, so all types of supplies—even eggs—may be handled by this system.

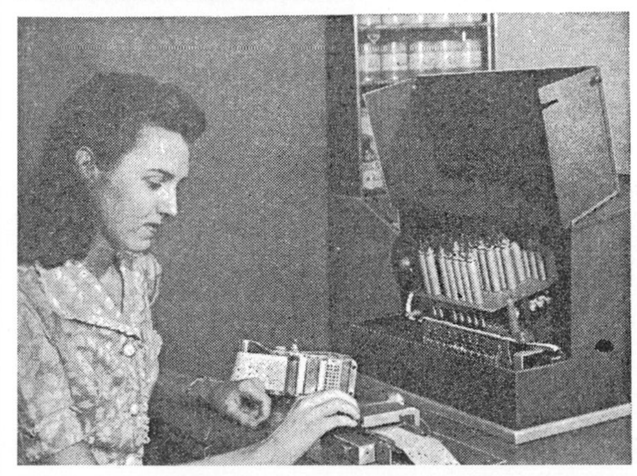

PREDICTION 1967

Ruth Sutherland programs a computer that can control the temperature, set clocks, and tabulate home bills.

ECHO with two of the front panels removed. The computer takes up about 20 square feet of basement floor space.

PREDICTION 1939

The combination room in the electric home of tomorrow contains a wide variety of electric equipment, operated from control centers throughout the house.

simple end-of-the-month routine.

ECHO will be programmed to keep track of real time so that events can be scheduled up to a year in advance with one-second accuracy. Ruth isn't interested in running their home on a second-by-second schedule, but she feels Jim won't be able to make excuses about forgetting birthdays and anniversaries ever again.

Recipes will be increased or decreased proportionately to provide any number of servings, with the necessary shopping lists printed out automatically. Later, as more complex programs are tried, the computer will generate balanced menus with specific calorie and nutrient content, from which the family can select their meals in advance.

ECHO provides the Sutherlands with a proving ground for experimental family games. The Sutherland children—Ann, 11; Sally, 7; and James Scott, 2—are looking forward to programming and playing games on the computer. As television displays are added to the system, many new games involving logic and strategy will provide family entertainment.

Jim says, "Computers are capable of being programmed to perform important household tasks today, but when we look ahead 20 years, even our wildest expectations will probably seem pale when compared to what ECHO, 1987 version, may be doing for us."

PREDICTION 1939 Manufacturers have come to look upon the design and distribution of home appliances as a long-term job of making electric homes. Today's house is a series of separate centers of electrification. Tomorrow's electric home will be build around the electric power supply and appliances.

This future home will probably be equipped with a number of control centers, from any one of which the homemaker can give her commands to appliances at work in the kitchen and laundry. Electric ranges already are equipped with automatic controls for temperature and cooking time, but there is no practical reason why these operations together with the other appliances cannot be controlled remotely from any room in the house. Perhaps short-wave radio may be utilized for this purpose, as well as for answering the doorbell and receiving visitors by transmitting a greeting to them and unlocking the door.

TRUE!

A Japanese inventor patented a bar code-reading remote control for microwave ovens back in 1989, and a simpler controller was patented in America in 2005. For some reason, the idea has never caught on.

3
MIND & WORD BECOME FAR-REACHING &
UNIVERSAL

A 1965 short story, "Dial 'F' for Frankenstein," began, "At 0150 GMT on December 1, 1975, every telephone in the world started to ring." An artificial intelligence was on the line to all humanity, born from the newly operational satellites orbiting the globe.

We now have an Internet with far more connections, but no spontaneous intelligence. Many who envisioned a world so deeply interconnected also had similar thoughts. The story's author, Arthur C. Clarke (who also wrote the film *2001: A Space Odyssey*), had proposed in 1945 that satellites in orbits of exactly one day would be ideal relays for all kinds of signals. He saw both the uses and misuses of a thoroughly linked world. So, too, did many who had earlier ideas of how to speedily connect us.

Radio was the most common agency, and countless inventors devised things like the "electric handshake" so people could meet and actually feel the hand of a stranger. That never took off, but in 1905 inventors started thinking about what we now call the fax machine, and electronic money transfer as well. As one bright-eyed article put it, "If that sounds fantastic, you should know that every device necessary for the accomplishment of this coming miracle is in operation today." In reality it took more than fifty years.

People wanted to connect—either with fast transport or communications. Radio was, along with airplanes, the hip, cool technology of the century. (A chain of radio magazines started the first science fiction magazines.) Worldwide mail delivery by "fast jet and rocket-propelled mail planes" got it half right, at least. "Radio delivery of facsimile newspapers directly into the home may be a reality in the near future" said a 1938 article—and they became available by 1990. *Popular Mechanics* got minor features right, too, predicting that push-button phones would replace dial phones—so nobody can actually dial F for Frankenstein today.

A 1950 prediction of "television telephone sends image of speaker picture-phone" was one of many that assumed we would want to have pictures along with sound on our telephones. We could now, of course, but there's no real demand. Instead, the software Skype makes this free over the Internet, anywhere in the world, even for conference calls.

People wanted pictures, sure, but for entertainment. The first public TV broadcast was of Adolf Hitler opening the Olympic games in Berlin in 1936, and commercial TV spread across the United States in the late 1940s.

Today, Web sites do rough language translations for free. Indeed, perhaps the most striking thing about these kinds of predictions is that people never thought so much information and service would be just given away.

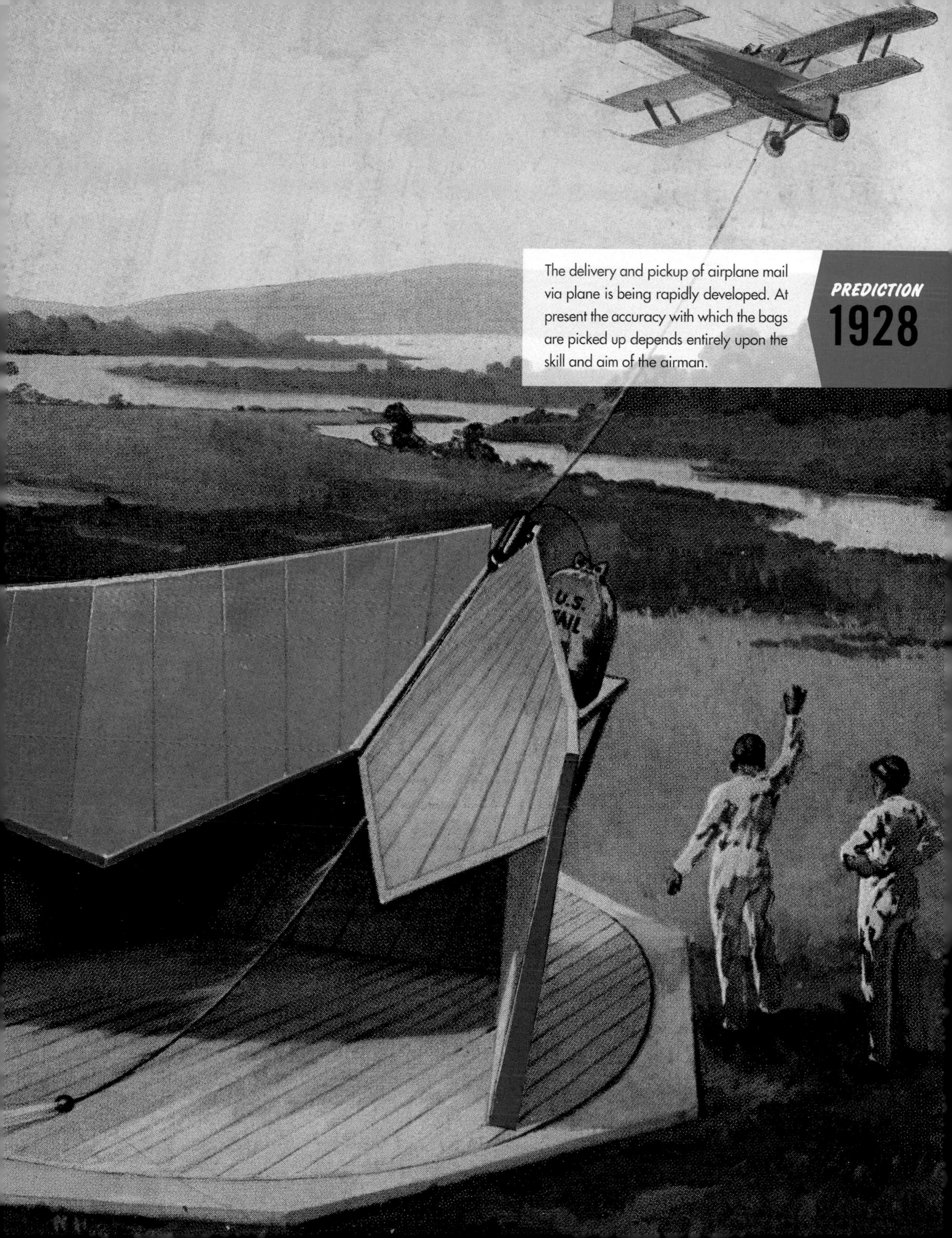

PREDICTION
1928

The delivery and pickup of airplane mail via plane is being rapidly developed. At present the accuracy with which the bags are picked up depends entirely upon the skill and aim of the airman.

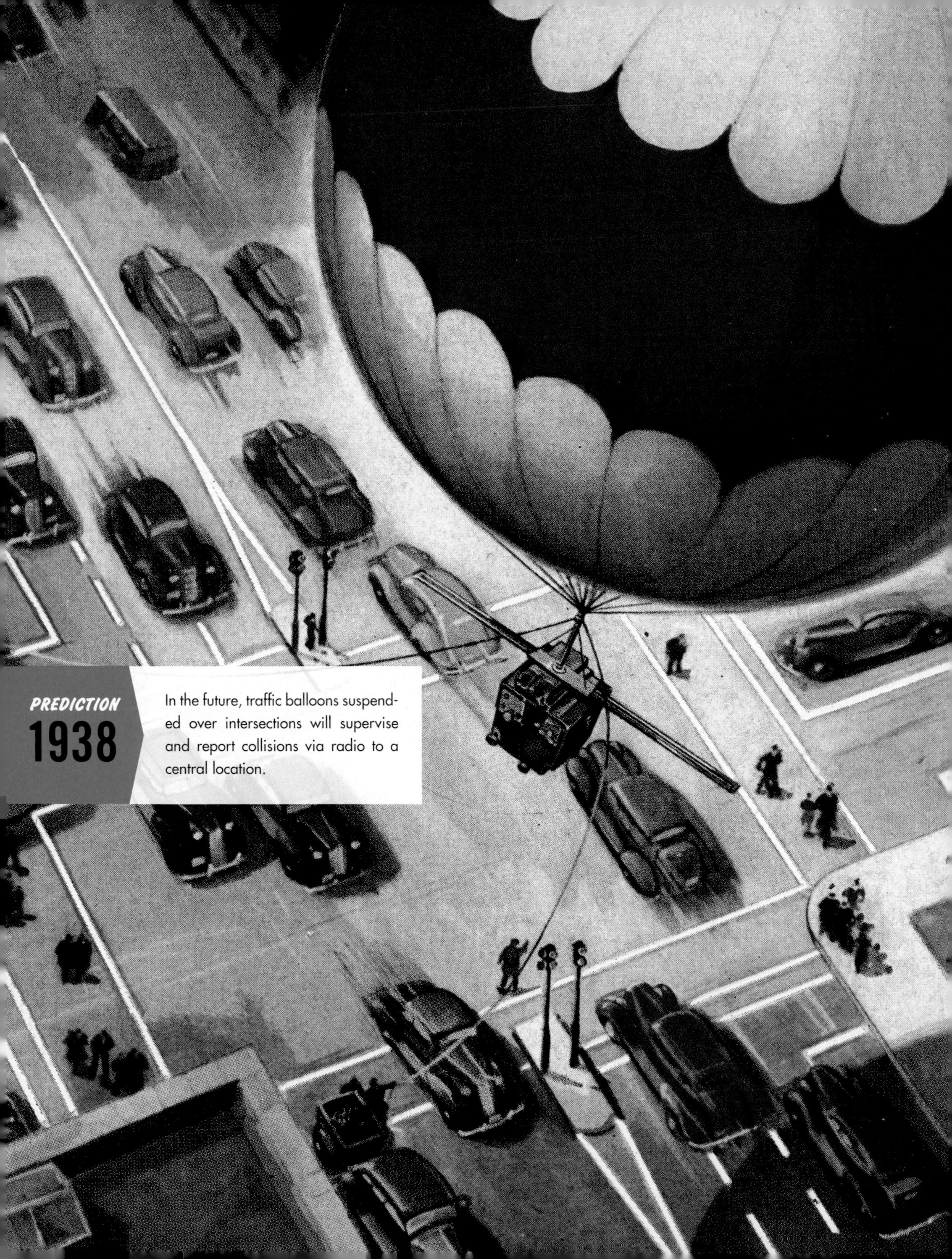

PREDICTION

1938

In the future, traffic balloons suspended over intersections will supervise and report collisions via radio to a central location.

Looking back on it all, it's useful to see how linear thinking can be outflanked by a wholly new idea. In the 1920s, contemplating that in the United States there might be as many as 50 million radio listeners, a pundit said, "The best solution may be a system of radio relay stations 20 miles apart on the level plains, perhaps 60 miles apart between mountain peaks."

Instead, satellites took over. Arthur Clarke's geosynchronous satellites now orbit by the thousands in what's now called the Clarke Orbit, doing this job for radio, telephones, and countless communications—at an astonishingly low true cost.

Push buttons are predicted to replace dial phones.

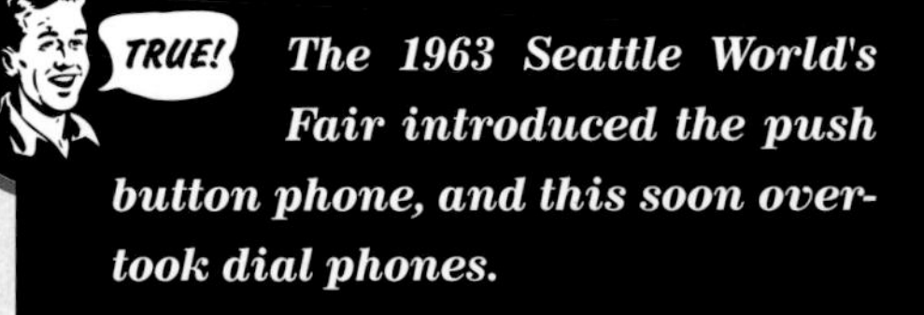

The 1963 Seattle World's Fair introduced the push button phone, and this soon overtook dial phones.

ELECTRONS SORT THE MAIL

PREDICTION 1905

From Chicago to Milwaukee in a straight line is 84½ miles. The fastest trains consume two hours in making the trip. It is now proposed to transmit mail and express matter between the two cities in 40 minutes via a pneumatic tube, 18 inches in diameter, conveying loads up to 500 pounds, with a series of 3-inch tubes for special express messages and very small packages.

Contracts for these long-distance pneumatic tube systems are now pending and their installation may begin in a day not far distant. There are now in operation in the United States more than 300 tube plants accommodating 6,000 stations, requiring 3,600 horsepower, and operating at a cost of $36,000 per day. The longest of these is the plant serving the Chicago post office. This system is nine miles long, double tubed all the way. It connects various railway and postal stations of Chicago with the old post office building and has a capacity for carrying 3,000 letters per minute each way.

Until very recently, however, the practicability of tube systems connecting cities hundreds of miles apart was precluded by the enormous amount of power which would be required to operate such a system. It is now claimed that a new system has done away with this objection and that a line connecting Chicago and Milwaukee capable of carrying packages up to 500 pounds in weight is an assured enterprise.

Instead of the carrier being forced through the tubes by means of high pressure behind it, as in most systems, the air is partly exhausted in front of the carrier, and the carrier glides along seeking to

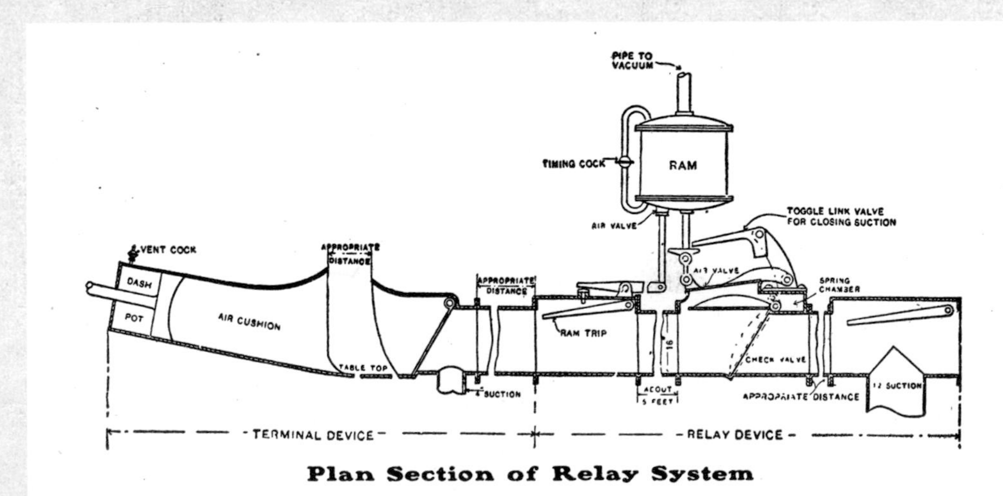

Plan Section of Relay System

The experimental plan for Chicago to Milwaukee's mail system will be capable of carrying packages of up to 500 pounds in weight.

MAIL DELIVERY BY *PARACHUTE*

Air-Mail Parachute Landing within a Few Feet of the Spot Aimed At: This Speaks Well for the Marksmanship of the Pilot. Lower Right: Dropping Mail Sack and Parachute by Hand, a Method Which is being Supplanted by Mechanical Release Devices

Above: Mail Bag and Parachute Suspended from the Bottom of an Airplane Fuselage. Below: The Case, Open and Upside Down, Showing the Parachute Carefully Folded. The Container is Stream-lined to Diminish Wind Resistance

The nonstop delivery of airplane mail via parachute is being rapidly developed in the United States, France, and England. Valuable matter—the only kind carried by airplanes—must be carefully guarded, which means, among other things, that it must be landed within a few feet of the person authorized to receive it. At present the accuracy with which the bags are landed depends entirely upon the skill and aim of the airman. However, some astonishingly close "hits" are being made with, and still greater accuracy is expected from, a two-speed parachute which is being developed in France. In the meantime it is quite safe to predict that parachute delivery will sometime become the rule.

5 **Each stack goes into another reader for a further breakdown into 16 smaller groups**

Envelopes are transferred to electronic reader which, upon instructions from electronic "brain," sorts coded mail into 16 stacks

6 **Eac**
is
are
are

Clerks operate Canada's new robot mail system.

demonstrate the old expounded by Newton—“Nature abhors a vacuum.” A system of relays will divide long pipe lines into sections of from two to three miles each, each section operating wholly independent of the others.

The carriers do not travel on wheels or rollers, but are covered with block felt which is as hard as rock, and fastened with brass caps and screws. It is expected the felt will last several months. The carrier is very necessary to the safe conveyance of the contents. A few weeks ago a leather pouch filled with mail fell into the tube at the Chicago post office and when the pouch arrived at the other end of the tube it was torn and riddled. This illustrates the tremendous power exerted by the air pressure.

An illustration of the Canadian Post Office’s robotic system.

PREDICTION 1955 Canada is now installing a robot system to sort the mail. Eventually the Canadian Post Office hopes to have a fully electronic system.

A clerk deposits the incoming mail in the hopper of a coding machine. The envelopes are fed automatically to a window in front of an operator. The operator glances at the handwritten address and punches keys on the machine to print the address in code on the back of the envelope. The envelope then goes to a stacker.

An operator carries a bundle of coded envelopes to the distributor, which sends them one by one through an electronic reader. The reader, upon instructions from the electronic brain, decides which of the 16 stacks is appropriate for each letter. These 16 stacks are then sorted into 16 additional subdivisions for 256 subdivisions total.

The Canadian Post Office visualizes for the future an electronic robot that will be able to read typewritten and some handwritten addresses. Meanwhile it will be testing its new robot, which has a much greater memory capacity than most human sorters.

HERE ARE RADIO'S LATEST NOVELTIES

PREDICTION 1924 Broadcasting radio programs simultaneously to 50,000,000 persons by means of a chain of relay stations so powerful that even inexpensive sets will be able to receive the messages, is a development that soon may be realized, an eastern scientist has predicted. From three to six stations would be necessary, according to this authority, one on the Pacific Coast, another on the Atlantic Coast, and a third in the central Southwest. Others might be erected in foreign countries, and all would operate on the same wave length to prevent overlapping.

PREDICTION 1924 During the previous presidential campaign, the insistence of party managers, and the natural desire of the people to look into the face and hear the voice of the two candidates, have almost compelled a candidate to tour the land from ocean to ocean on a speaking trip, even when he was our President in office at the time. However satisfactory this has been to the voters, such an effort must be at great cost in health and vitality to the candidate, and in the case of one at the time occupying the highest office in our land, with a certain loss in dignity, coming down, as it were, to the level of stump orators and candidates for ordinary town and county offices.

The radio, however, avoids this condition, and the utterance of a President from the White House carries dignity and weight, even though his utterance is really in furtherance of his own re-election. We predict that in years to come this method of appeal will be found acceptable to voters, campaign managers, and especially to the candidate himself.

Tomorrow's farmer, if a diorama at the New York World's Fair prophesies are right, will be rather a radio navigator and dispatcher than a tiller of the soil. The diorama depicts the farm engineer seated in his glass-enclosed radio tower, manipulating switches that transmit radio signals to the robot machines that sow and cultivate, irrigate, harvest, sort and can, freeze and pack the crop, all by remote control.

PREDICTION
1926

Recent developments in radio include airplanes guided by radio beacons and radioed weather reports, portable "parasol" aerials, and a low-power short-wave system designed to survive even the most perilous of expeditions.

Besides, there is a strange, very forceful psychology, when a listener in a small remote hamlet realizes the words he hears are at the same instant being received at every other hamlet and city from the Atlantic to the Pacific, and from the Great Lakes to the Gulf, and even our possessions beyond the seas. It is a joy to be privileged to live in an age when mind and word are becoming so far reaching and universal, and which will very soon extend their utterances to the farthermost point on our Earth.

PREDICTION 1926 Dr. Lee de Forest, "Father of Radio," visions tremendous developments in radio in the next few years. "Broadcasting will take on entirely new phases and conceivably the national government may take over all broadcasting of a general or public nature. We shall have radio schools—a whole system of education in which all instruction is given over the radio, and we shall have radio newspapers and magazines, and perhaps a radio church.

"To me the delightful thing about radio is that it enters so indispensably into modern life. It extends the hearing power of the human ear around the earth, or will do so, to enable anyone to listen to audible sounds anywhere else. Some day in the future man may even go out from his own planet and beyond his own atmosphere and if other planets are inhabited by sentient beings, he and they may find it possible to communicate. But not by radio as we know it today.

"While it has already made itself indispensable in our daily lives, it is a fact that radio is yet in its merest infancy."

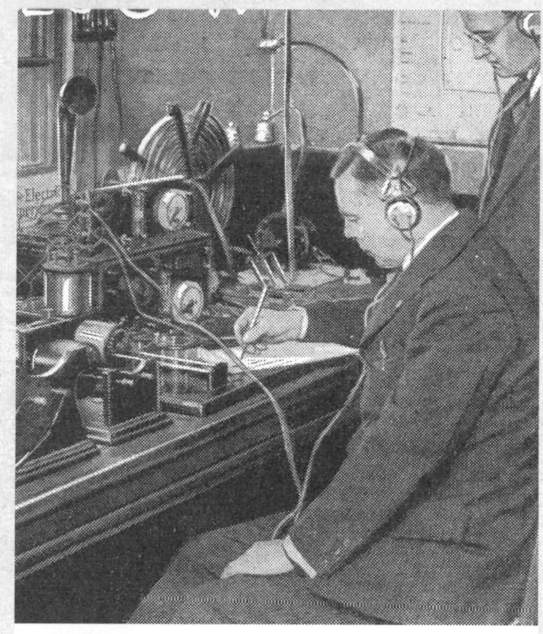

PREDICTION 1926 Dr. de Forest at a modern radio set. How swiftly radio has changed in recent years!

PREDICTION 1938 Radio delivery of facsimile newspapers directly into the home may be a reality in the near future. Only perfection of certain technical details stand in its way, according to radio experts. The perfected system will permit reception of the newspaper on a machine during the night. Present equipment transmits and receives pages the size of the standard letterhead. The next morning, the owner takes the sheets out and reads his newspaper.

PREDICTION 1938

Interior of facsimile newspaper receiver, showing sheet of news received by the facsimile method.

PREDICTION 1950 By A.D. 2000, fast jet and rocket-propelled mail planes will make it so hard for telegraph companies all over the world to compete with the postal service that dormant facsimile-transmission systems will be revived. It takes no more than a minute to transmit and receive in facsimile a five-page letter on paper of the usual business size. In this future city, the clerks in telegraph offices no longer print out illegible words. Everything is transmitted by photo-telegraphy exactly as it is written—illegible spelling, blots, smudges and all. Mistakes are the sender's, never the telegraph company's.

RADIO MOTION PICTURES IN THE HOME

1923 In a recent test before government scientists, pictures of a moving hand and other objects were transmitted by radio. The reproduction was said to have been somewhat indistinct but unmistakable. Improvements now in course of completion will bring out sharply the transmitted "movie," and the prediction was made that the near future will see radio motion pictures in the home become one of the world's most popular pastimes. The distance to which the pictures may be sent is said to be limited only by the capacity of the sending station.

SIMPLE IN *ARRANGEMENT* AND SOFT IN COLOR

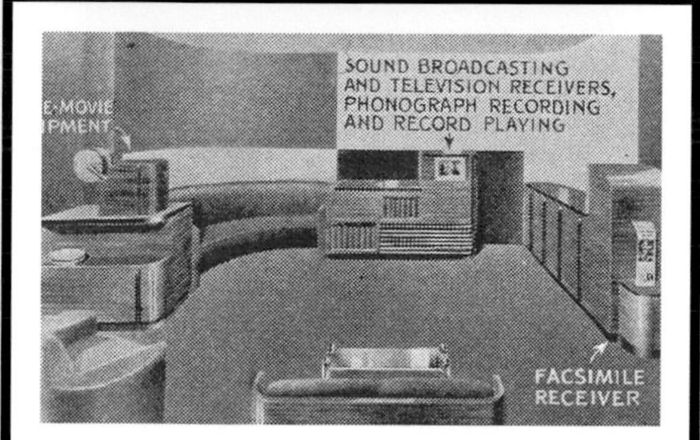

The "radio living room of tomorrow" includes various sight, sound, and facsimile facilities.

"But is this true television? By that term we really mean the ability ourselves to see over long distances. Not to be shown pictures taken far away, but to actually view the scene itself in Los Angeles while we are in New York.

"Can we ever do so fantastic a thing? Theoretically, it is possible. Commercially and financially, it is impractical."

PREDICTION 1926

"Television we have now, in a way," says Dr. Lee de Forest, inventor of the radio tube known as the audion. "By using the photo-electric cell, we can transmit pictures and writing, either by wire or by air. This will become commonplace.

"It is even said that a coronation in London may be pictured in New York theaters the same evening. Before such a feat is possible, however, coronations may have gone out of fashion. Yet the transmission of still photographs is already underway and sending motion pictures is but a matter of working out technical details.

Television studio of tomorrow may become a reality after Victory Day. Americans can look forward to airborne talking pictures mirrored to you from a 12-inch cathode ray tube.

PREDICTION 1944

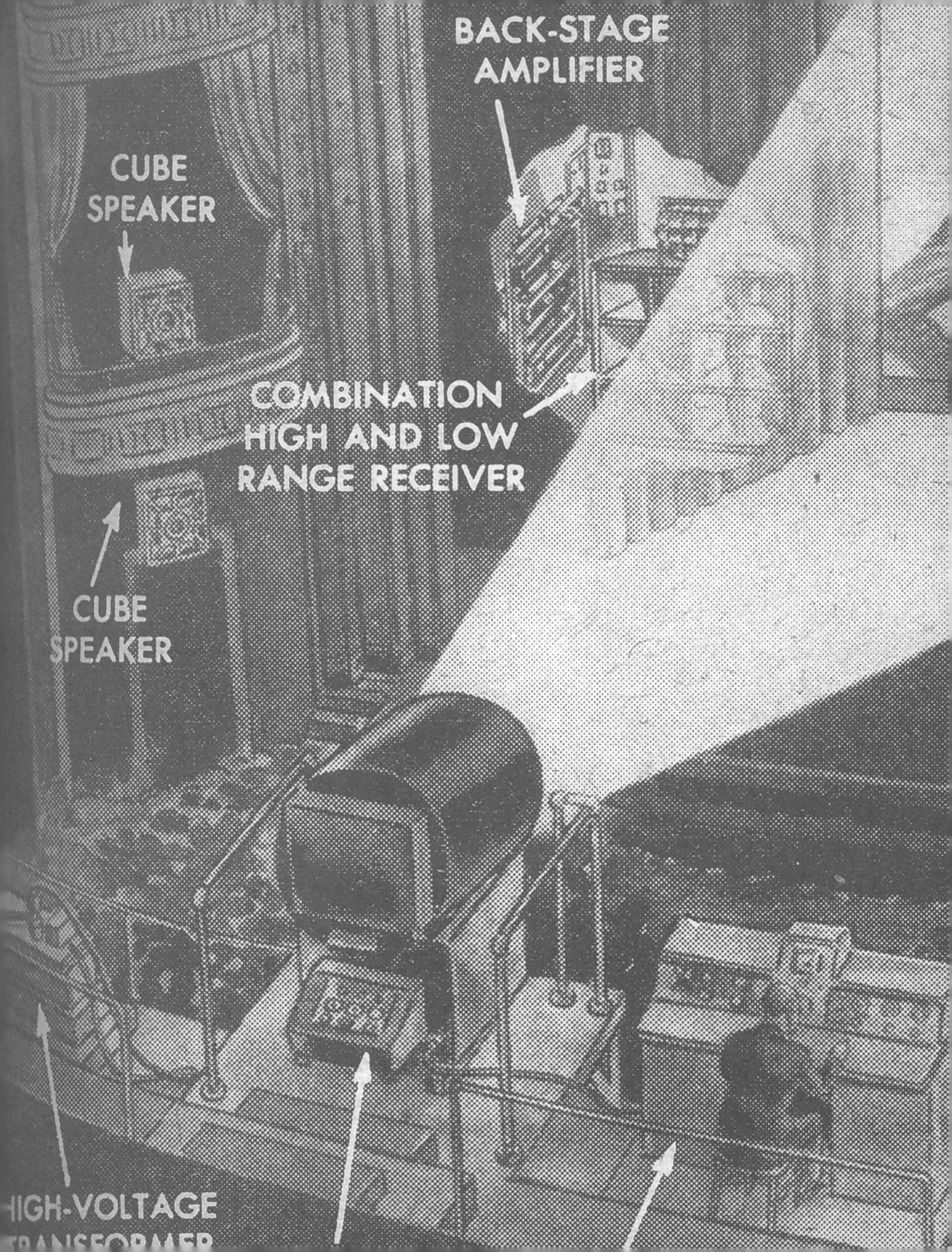
BACK-STAGE
AMPLIFIER
CUBE
SPEAKER
COMBINATION
HIGH AND LOW
RANGE RECEIVER
CUBE
SPEAKER

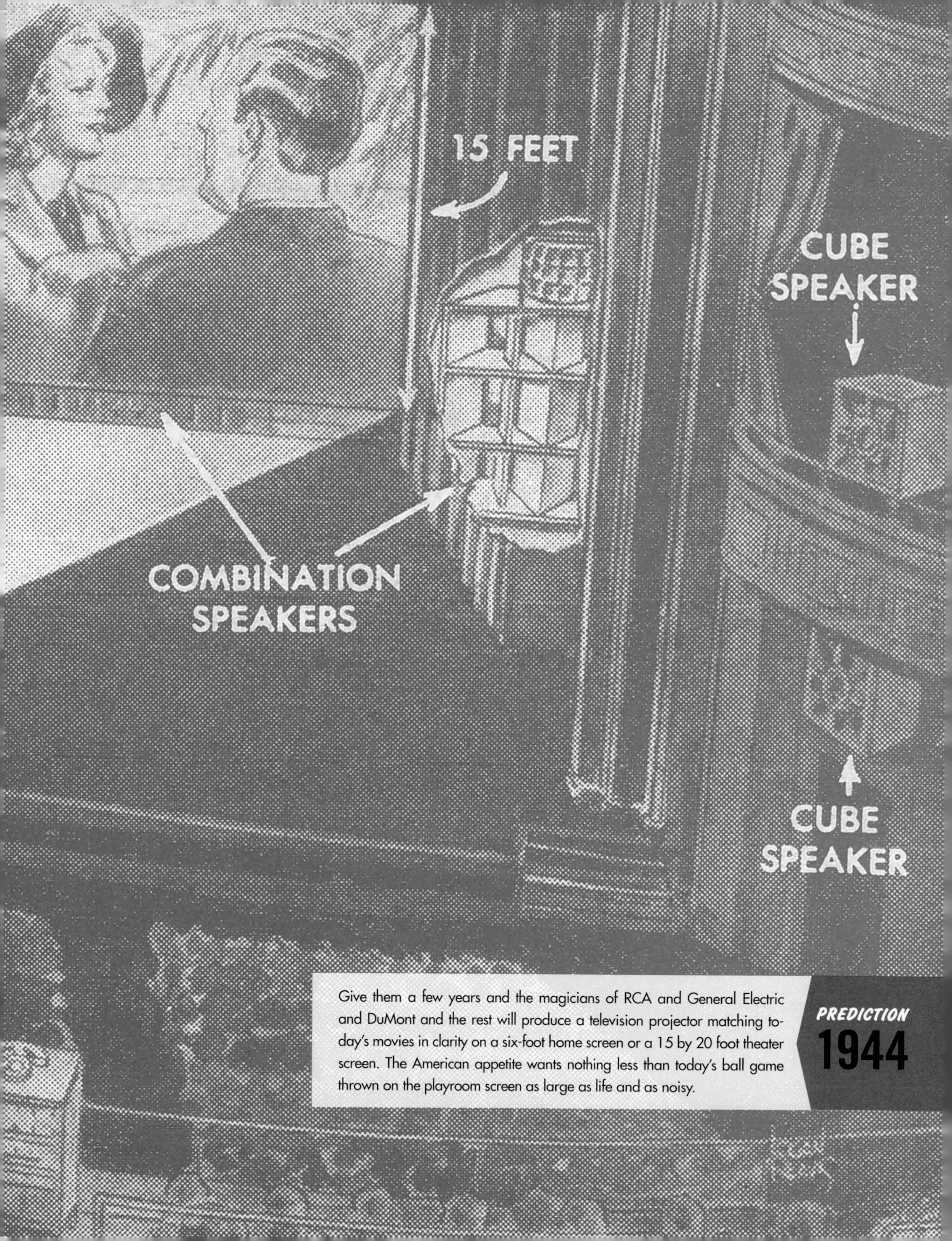

15 FEET

CUBE SPEAKER

COMBINATION SPEAKERS

CUBE SPEAKER

Give them a few years and the magicians of RCA and General Electric and DuMont and the rest will produce a television projector matching today's movies in clarity on a six-foot home screen or a 15 by 20 foot theater screen. The American appetite wants nothing less than today's ball game thrown on the playroom screen as large as life and as noisy.

PREDICTION

1944

PREDICTION 1944 Look a bit into our future with us. On your living room wall a televised Hollywood movie plays for a roomful of guests. It is televised in full color, and you are boasting that any day now there'll be "airborne" movies in three dimensions.

One of your guests is ready to go home, and you telephone for a taxicab. In a few minutes it arrives, summoned by shortwave radio while cruising. Before retiring you turn on your facsimile set to record the morning news while you sleep; and suddenly your FM receiver automatically comes alive with an important news flash from a master transmitting station that can call in your set without your own help. While you were out, the magnetic wire recorder took over as "home secretary" to record any phone calls during your absence.

RCA laboratories have built experimental sets with an 18 by 24-inch translucent television screen for home use.

Its growth stunted by the war, television is enduring a painful adolescence. What's ready, if Victory Day were tomorrow, is the promising if primitive marvel that went on the air to the New York, Schenectady and Hollywood neighborhoods five years ago: airborne talking pictures mirrored to you from a 12-inch cathode ray tube, at times flickering and lightstruck and distorted like a 1910 nickel show. What's ahead is—super.

Television engineers promise you that true color television can be produced 10 years after you relieve them of the burden of war research. A mechanical method of televising color with whirling disks synchronized to a related mechanism in the home receiver has been tried; but electronic color is what the broadcasters order, and it will come.

An expensive problem remains: the transmission of FM sound and television image beyond the horizon. The short waves don't follow the earth's curve. A nationwide hookup involves transmission from city to city by coaxial cable or by radio relays. Either method is costly. The American Telephone and Telegraph Company considers as a definite possibility the investment of up to $100,000,000

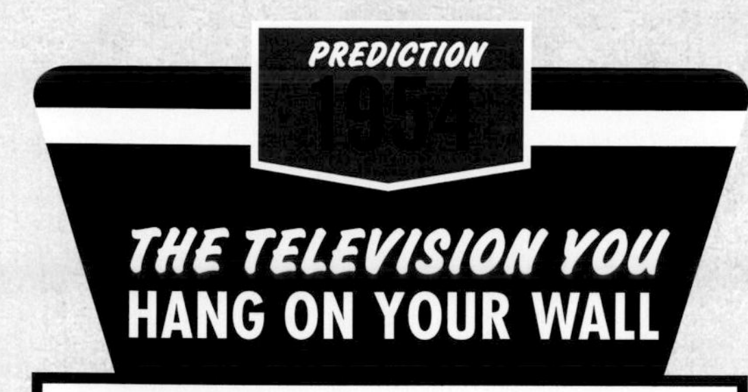

THE TELEVISION YOU HANG ON YOUR WALL

General Electric scientists predict your TV-picture screen in 1964 may be so thin that it can be hung like a painting on the wall or mounted like a vanity mirror in a table model.

TRUE! ***Today's LCDs hang on walls around the country, but in 1964, the first plasma displays were just beginning to be developed, and still encased in unwieldy boxes.***

in construction of 6,000 to 7,000 miles of coaxial cables which could carry television images as well as hundreds of telephone conversations. Even that mileage would serve only metropolitan cities, and television transmitters can broadcast no farther than they can see: 100 miles is a limit seldom reached.

Television will come to your home. Figures do not stagger the men who believe theirs is the great new industry of tomorrow. The frontier once was measured by acres, valleys, states. Today it is measured by opportunities. Television is one of our greatest frontiers. When the brains that developed radar and all its secret electronic affinities are turned loose on FM and AM and television, that new frontier will be wide open.

RADIO COMMUNICATION ENGINEERS SHRINK THE GLOBE

PREDICTION 1944

While airplane designers toil night and day for blinding speed, another group of scientists is just as busy working to place the whole world at your elbow. These are the radio communication engineers. Suppose you let them take you for an electronic ride into the future.

Seat yourself in an easy chair in the parlor, cross your carpet slippers, and settle back to read the evening newspaper reproduced page by page on your television set. A bell rings in a box at your elbow. Someone hurries into the room and says that it must be a letter from Johnny, so you flick a switch on the facsimile machine and put in a piece of paper. A cylinder soon shuttles back and forth, turning slightly at the end of each movement. A tube sprays little ink spots occasionally onto the paper. It's Johnny's handwriting all right.

If that sounds fantastic, you should know that every device necessary for the accomplishment of this coming miracle is in operation today. Only expansion and a market are needed.

PREDICTION 1944 On the radiophoto machine, a photograph can be reduced in size by photography to 6½ inches wide, and then radioed to Moscow, 4,614 miles away, in about 13 minutes, after which they are enlarged to original size. An airplane would have to travel approximately 21,300 miles an hour to get them there as quickly.

Or the pictures, reduced to an area of five by seven inches, will reach Melbourne, Australia, 7,420 miles from San Francisco, in just 10 minutes. The elapsed time changes only with the short width of the transmitted picture, since it requires two minutes for the magic eye of the machine to scan a linear inch, starting from the top. The difference in distance in sending to Moscow and Melbourne isn't much since the radio impulses circle the earth something like seven times in a second—a record for airplanes to shoot at for some time.

Photos and facsimiles can also be exchanged between New York and such distant points as London, Cairo, Buenos Aires, Berne and Stockholm. Berlin and Tokyo are out of service at this time. Many of the war pictures which keep Americans the best informed people on earth have been sent in this manner.

Not long ago, Mrs. Gerald Mayer of New York City wrote her husband, whom she had not seen for a year and a half, saying she wondered how he was looking. In reply, he radioed a specially posed picture. Mrs. Mayer then radioed hers to him. It reached Switzerland in 10 minutes.

You'll probably be hearing more of the accomplishments in radio communication when peace lifts the security veil and you will become famous if you can figure

PREDICTION 1905

ELECTRIC *HANDSHAKE*

To discern every expression on the face of the one you are talking with, to hear his voice and feel the pressure of his hand, when separated by hundreds of miles, is the ambitious prediction of French scientists. Under such circumstances the physician could safely prescribe for a patient in another city.

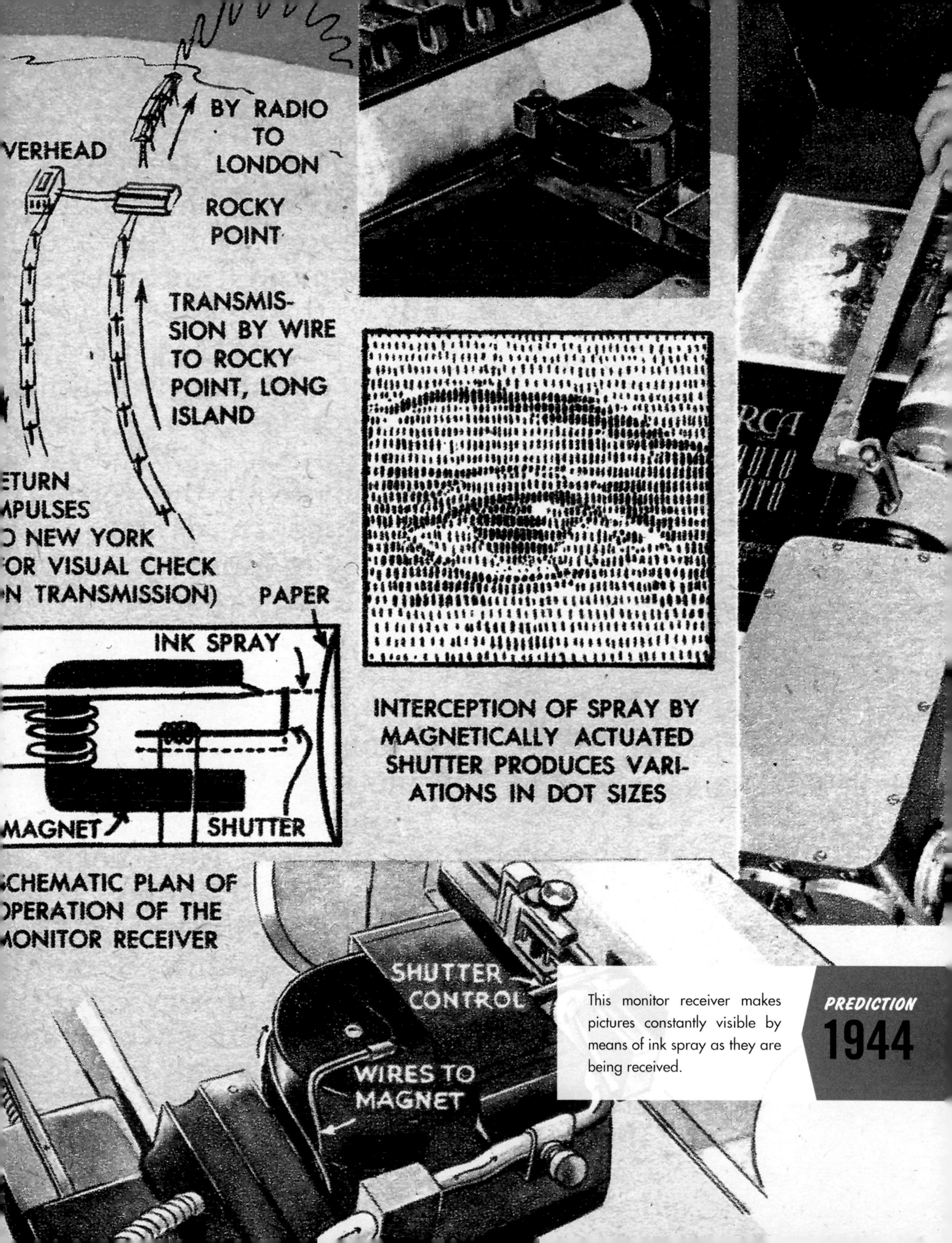

SCHEMATIC PLAN OF
OPERATION OF THE
MONITOR RECEIVER

INTERCEPTION OF SPRAY BY
MAGNETICALLY ACTUATED
SHUTTER PRODUCES VARI-
ATIONS IN DOT SIZES

This monitor receiver makes pictures constantly visible by means of ink spray as they are being received.

TELEVISION SENDS IMAGE *OF SPEAKER*

Still far from practical realization, this apparatus combines a portable television transmitter with a push-button telephone. When the receiver is lifted, the image of the person making the call flashes on the screen. Fantastic as it may appear today, engineers believe that television telephones may become commonplace within a generation.

TRUE!

Video telephones were developed in the 1960s, but they have never achieved widespread adoption.

out how to send television waves without relay towers on land or sea. These high frequency waves have a stubborn way of refusing to follow the curvature of the earth, flying off into space. Some scientists say it can't be done, but they said the same thing many years ago when a pitcher threw the first curve with a baseball.

COMPUTERS SPEAK THEIR MINDS

PREDICTION 1954

Russian is translated into english by an electronic "brain" at Georgetown University that normally busies itself with nuclear physics, trajectory-plotting and weather forecasting. A girl who doesn't know a word of Russian punches out the foreign words on a standard IBM machine. The punched IBM cards are then placed in a reading unit. Guided by six basic rules of syntax and grammar, the "brain" searches through its 250-word vocabulary of English equivalents, frames them into a sentence and prints the translation seconds later. Code numbers to various words govern necessary changes like reversing word order, choosing between two meanings and adding connectives. Within five years, it is predicted, such a system will be practical for translation of several languages.

PREDICTION 1959 **Library machines that look up references for you and summarize their contents** in as many words as you wish are forecast within 20 years. Such an information-retrieval machine would hunt through memory drums on which the complete texts of millions of documents might be stored. It would set aside all references to a desired subject, then assemble

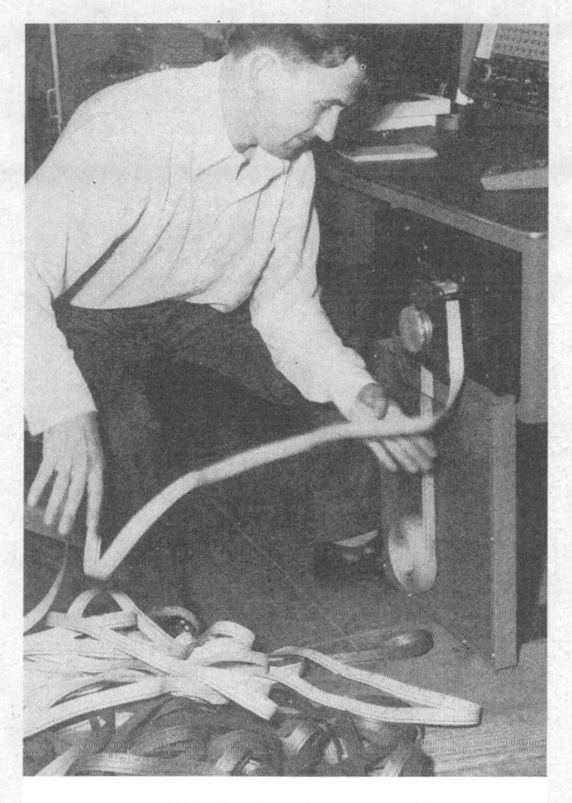

This drawing shows a Michigan inventor's translating machine created in 1958 employing paper tape. The interpreting telephone was first predicted in 1910.

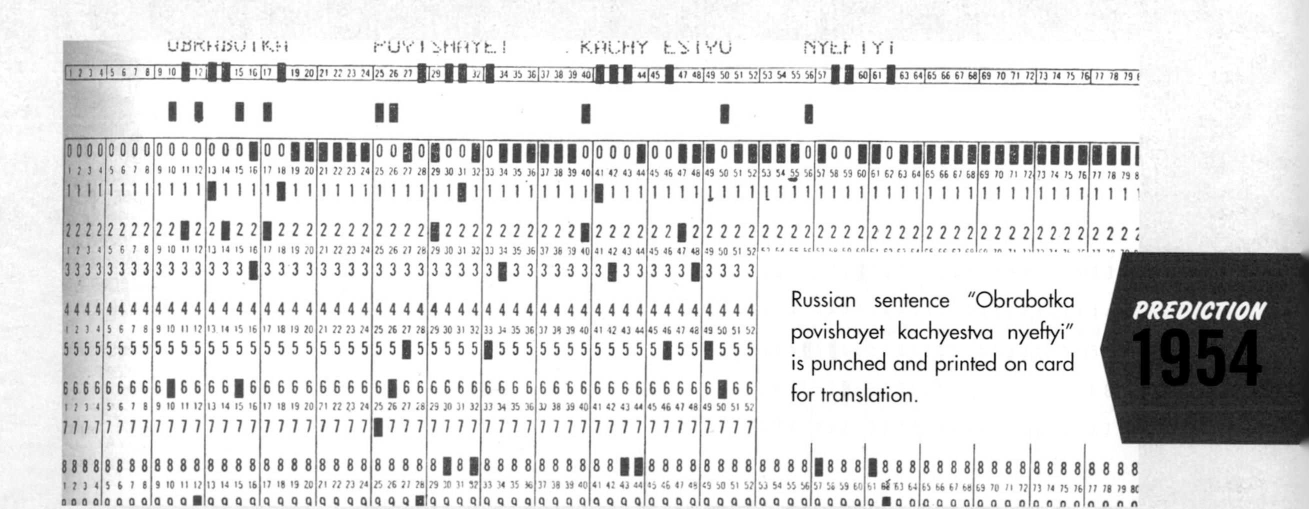

Russian sentence "Obrabotka povishayet kachyestva nyeftyi" is punched and printed on card for translation.

BASIC SPRING-DRIVE WATCH

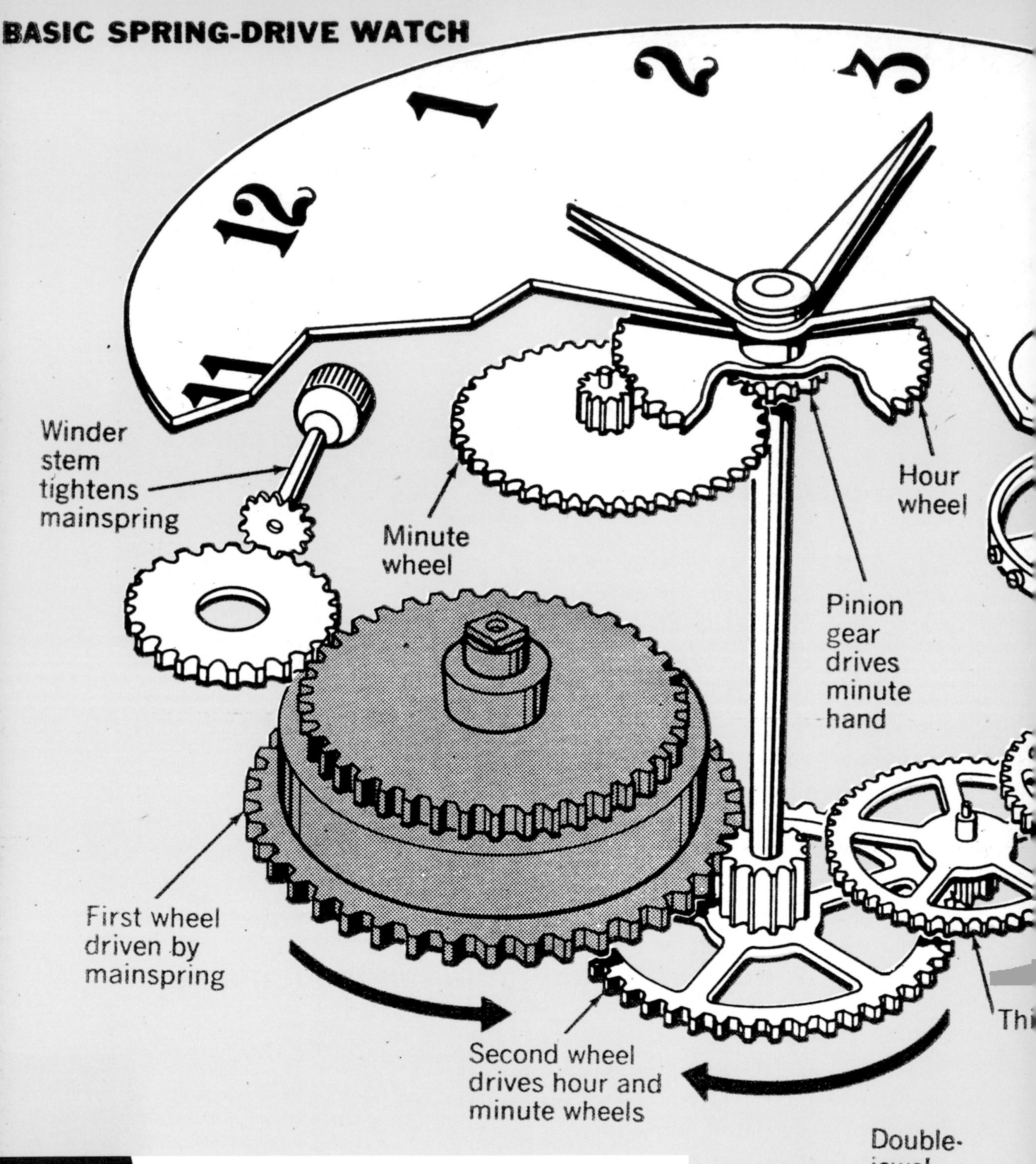

PREDICTION

Waltham engineers foresee this exciting possibility: "Wristwatches in the year 2000 will be used for more than time measurement. They will be total communications centers, containing devices not only for accurate timing but also for voice and vision communication and recording—they'll even contain simple miniaturized computers." How about that for a prediction!

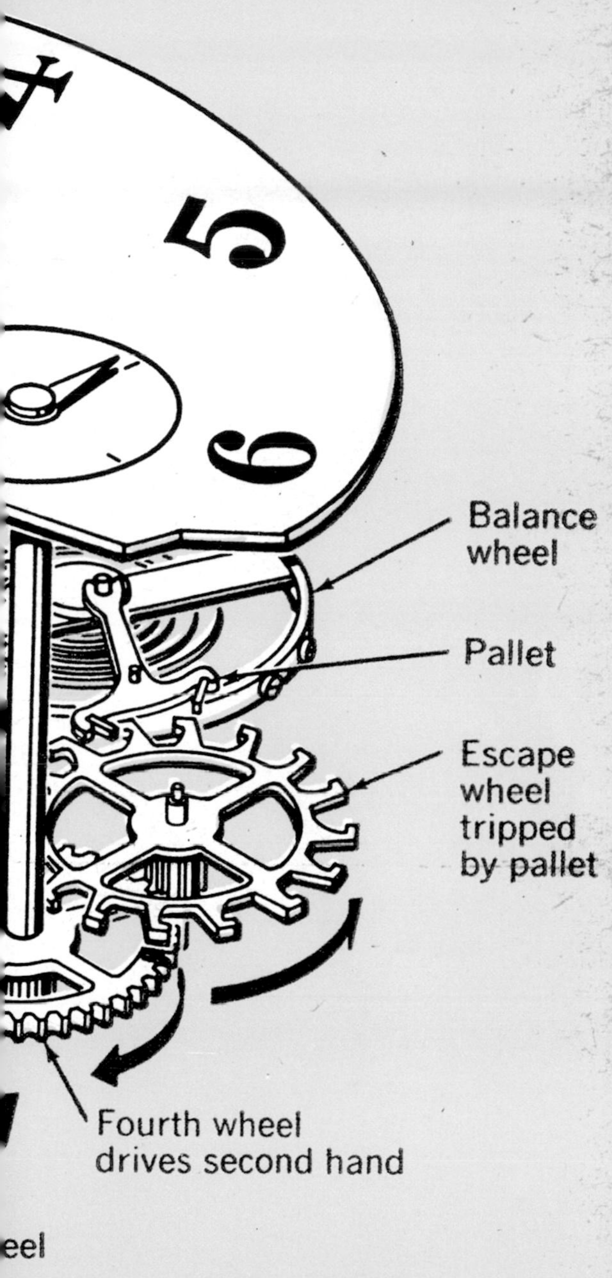

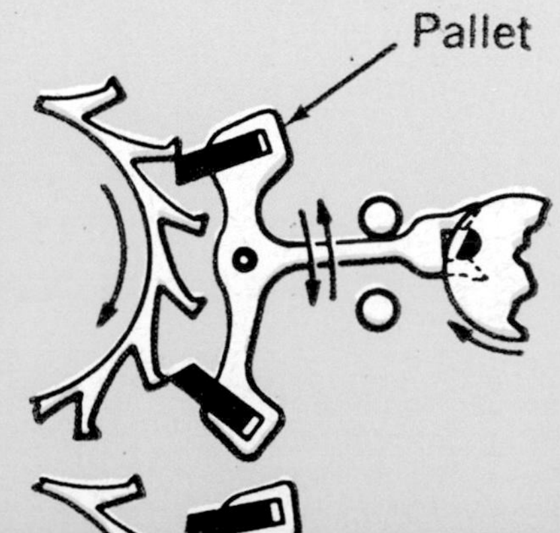

PREDICTION 1962 Businessmen may soon be able to carry computers around in their pockets to make lightning-fast calculations while away from their desks or offices.

The prediction is based on the recent development of a portable computer the size of a small suitcase, 18 by 12 by 20 inches, which can make calculations previously available only on giant computers.

The first pocket calculators were developed in 1970.

the facts from each reference and present them in smooth sentences. The chief stumbling block to such machines at present is that humans must assign subject headings or "descriptors" to the masses of material. It is expected that this rudimentary form of translation machine can be perfected long before machines are developed that can make good idiomatic translations.

PREDICTION 1962 When an electronic computer of the future feels it is being overworked, it might just speak up and say, "Take it easy, boss, my transistors are aching something awful."

Three scientists at Massachusetts Institute of Technology have developed a synthetic speech machine that can sing simple songs or say a few sentences.

4
HEAVY WATER MAY PROLONG
LIFE

PREDICTION 1949

The 20th-century '49er is looking for uranium—not gold. He carries a radiation counter that reveals radioactivity in rock. In this issue of ***Popular Mechanics***, we show you how to build your own lightweight portable radiation detector usable for prospecting uranium.

Everybody wants to live longer. The average American in 1900 lived to about fifty; today the average age is around seventy-eight. Yet few remark on this 50 percent increase!

Instead, our lives adjust imperceptibly. The young today marry later and seem in no big rush into adult life. Remarkably, the same 50 percent increase also happened in the 1800s. Perhaps the most striking aspect of these big, dramatic improvements in life spans is that they have derived mostly from improvements in the survival rates of children. People have gotten to the middle of the age distribution—say from forty to sixty—mostly through vaccines, cures, and improved medical technology, especially computer-based technology of late. Still, doctors relying exclusively on the diagnostic capabilities of computers is not nearly as widespread a phenomenon as many once expected. Human judgment still plays a major role, and doctors never like to be sidelined.

Many new tools—ultrasound, lasers, CAT scans—have helped. And the elderly death rate has shown some improvement—but not a lot. There is still a fairly solid "wall" around age eighty, and beyond it, the population declines roughly exponentially. One might term this the "fragility wall," where people become prey to any passing microbe or severe accident. Their resistance and resilience erodes until they are easy marks.

Also, consider the below figure. Notably, since 1900, when the death rates of the sexes were just about the same, women's longevity has consistently outpaced men's; today they live about 10 percent longer in all advanced societies. Few take note of this remarkable inequality, which is still increasing after a century. Females consume more than two-thirds of health-care budgets and are consistently heavier users of health services throughout their lives. Upgrading male longevity to the level of females' in the advanced societies would improve the average human survival more than, for example, completely eliminating cancer. This strongly suggests that social forces beyond the reach of technologies alone have a great deal to do with improving our expected life spans.

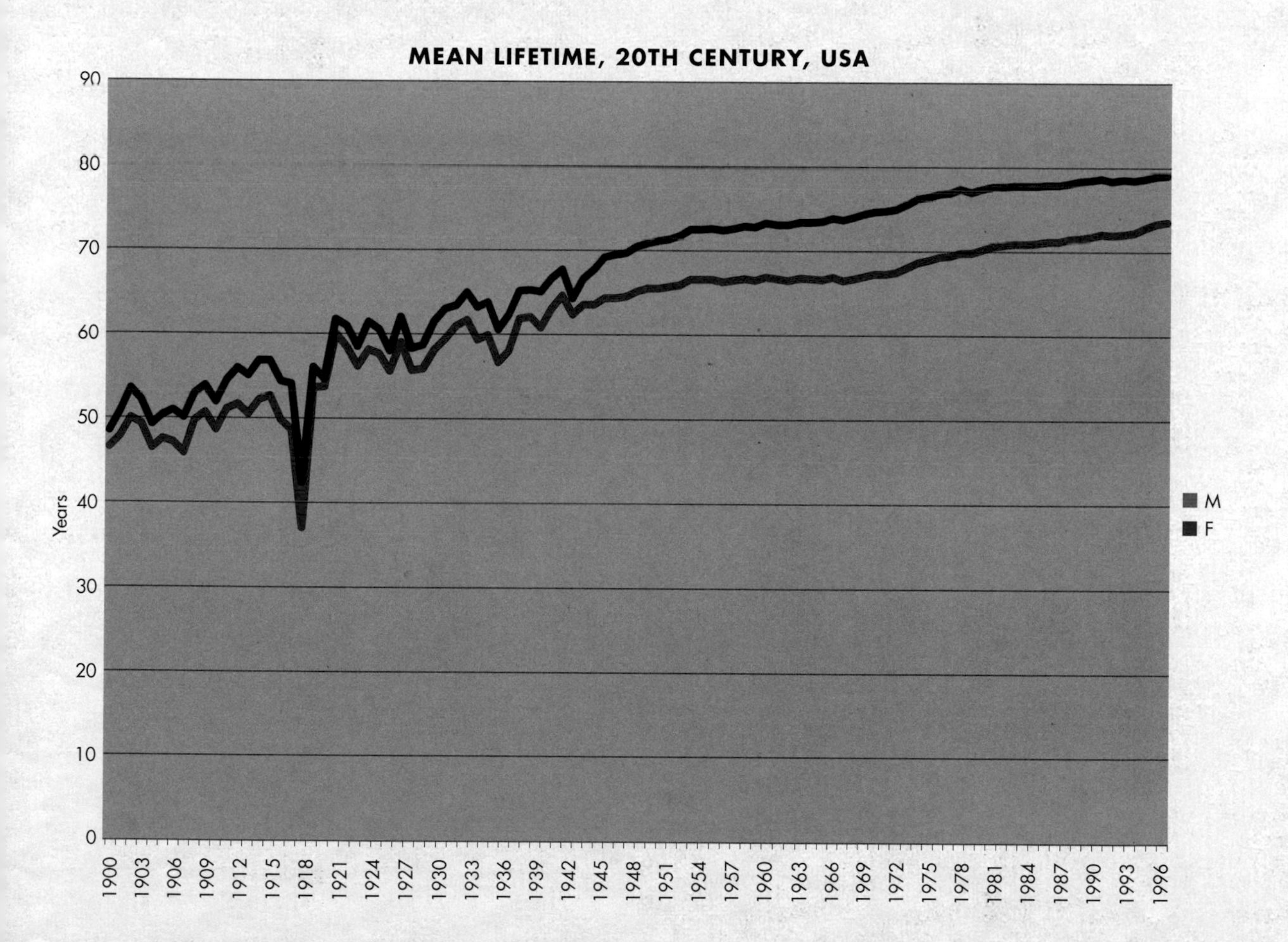

In addition, such extensions begin to change our views of the human condition. We already see that young people are delaying education, marriage, and other social goals longer than what was typical in the first half of the twentieth century. This may be come from their sense that they have plenty of time, since they see their twenty-first-century parents leading vigorous lives well into their seventies and even their eighties, a phenomena nearly unknown only a century ago. Such subtle changes go unnoticed because they are slow and intuitive. For example, the Social Security retirement age was set at sixty-five in the 1930s because half the population would be dead by then, and would therefore never collect!

Perhaps the biggest factor shortening life spans in the early twentieth century was simple ignorance. Before 1950, few realized that the modern healthy lifestyle—no smoking, eating plenty of vegetables and fruits, avoiding fats—was solidly grounded. The greatest achievement of the twentieth century was to prove these simple guidelines in clinical trials.

How far can this go? We have no true idea of an upper limit on life span, though the oldest person known was a Frenchwoman who once sold pencils to van Gogh and died at age 122. If we eliminated all aging, so that we faced no "fragility wall," if we completely eliminated diseases and could avoid all causes of death except accident (including suicide), how long could we live? Most people, when asked, guess at ages like 120 or 150. The answer, gathered from studying the causes of death in actuarial rate tables, is astonishing: close to 1,500 years!

With only a century or less of life, humans have developed many social forms to deal with this life span, and nearly none that look beyond it. Take just a small step into that immensity: imagine living to 150. How would you plan a career? Could you keep interested, if the job (like most) had a fair level of routine? And what about marriage? Some argue that the divorce rate is high these days because people know they face a far longer life span together than a century ago. Perhaps marriage itself can be redefined to set term limits, an idea which was called "contract marriage" in the 1940s and never caught on. And what of children? As we live longer, the population growth problem worsens if we keep reproducing. Our lives will be less family-centric because our children will be adults, with their own lives, for a larger fraction of our spans.

With augmented bodies and cures, the promise of the late twentieth century is still being fulfilled, in steps. If that simple life span improvement of 50 percent per century holds up, we will live on average to around 120—the previous world record. ◎

A flying ambulance, which is an ordinary motor car on the ground, but can take place in a plane and become the body of a ship when it takes to the air, has been designed by M. P. Delcourt of the Eiffel Laboratories.

PREDICTION

1928

PERIODIC CHECKUPS BY COMPUTERS

PREDICTION 1957 Honeywell company president Paul B. Wishart recently asked several of his key scientists to set forth in memos what they thought life would be like in A.D. 2000. Among other answers—not "blue-sky stuff," I am told, but rather conservative predictions from highly regarded members of a conservative profession—they suggest that machines will diagnose and, in many cases, treat human illness. Since sickness usually produces chemical changes in the body before severe onset, frequent, periodic checkups by these machines will make possible the prevention of most diseases.

PREDICTION 1966 "What is the diagnosis, computer?" When your family physician first asks this question of his latest diagnostic tool, you may shudder and long for the good ol' days of bedside visits and the personal touch in medicine. But with the latest medical findings computer-stored at your doctor's fingertips, you can do nothing but get well faster!

PREDICTION 1966 Physicians prepare symptom cards for computer input at the American Medical Association convention. Readout of diagnosis, flashed on the screen, is almost instantaneous.

THE AMAZING NEW COLD AIR TREATMENT

PREDICTION 1905

In a sultry day in August in a Southern city, men drop in the streets by scores, overcome by the terrible heat. Quite soon in the future, if you should follow one of those senseless, dying bodies through the hospital doors, you would see it laid out on a wheeled stretcher, rolled along a hallway. The patient will pass through several tight fitting doors which automatically open and close, and be carried into a small room where the air is like that on a snow clad mountain peak: cool, clear, absolutely dry, and fairly sparkling with vitality.

Cold air treatment as described does not yet exist, but a plant has been in successful operation for several months producing unlimited quantities, as desired, of either moist or absolutely dry air at extremely low temperatures. Less than a year hence, this treatment will doubtless cause no more wonder than other great discoveries, and like the X-ray and wireless telegraph, pass out of the realm of mystery.

If cold storage is desired, that can economically be produced. There has developed thus far no apparent reason why the cold, dry air, at any temperature desired, cannot be delivered to residences, stores, hotels, etc., in identically the same way that gas and water are now delivered. Bedrooms, or sick rooms, can be kept cold by running a rubber pipe extension.

A remarkable feature in the cold room experiments has been that the operators and others have repeatedly gone into the cold room where the temperature was at zero, while in the shade outside it was around 90 degrees, when they were

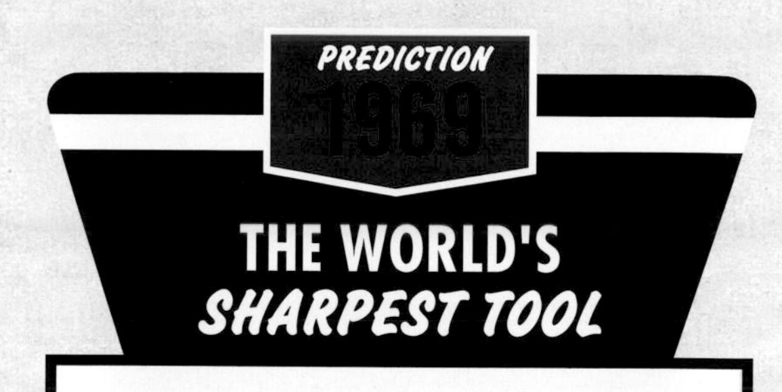

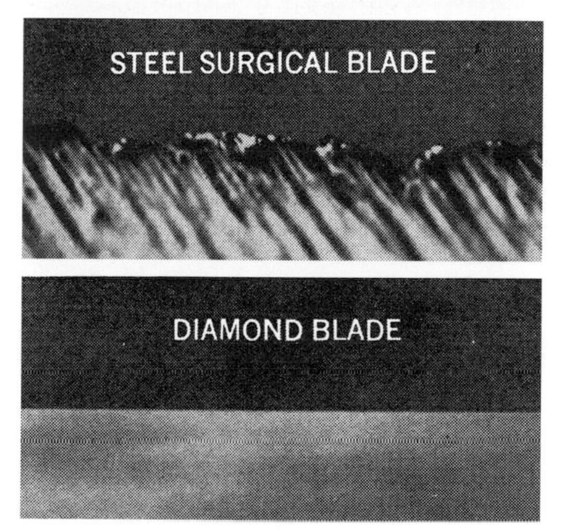

The blade in this knife was made from a diamond of gem quality and has a cutting edge the width of 12 carbon atoms—approximately 1/1000 the width of a human red blood cell. At 1000-X magnification, the edge of the diamond blade is a sharp unbroken line while that of the best surgical steel blade looks as jagged as the Alps. Ophthalmologist Dr. Davis G. Durham has used it to perform over 200 eye operations and says it would be excellent for operations in the inner ear or for blood-vessel grafts.

perspiring at every pore and so far none have taken a cold. The explanation is the absolute dryness of the air. This has convinced several prominent medical experts who have witnessed the demonstration that the cold, dry air process will be the successful treatment for tuberculosis and pneumonia, as well as typhoid, scarlet, yellow and other fevers and diseases.

A tuberculosis specialist recently announced the plan of taking a colony of patients to Greenland, that they might breathe the cold, dry air of the Northland. The remarkable health of Arctic explorers, in spite of manifold hardships and terrible exposures, is well known. There is no reason now why we cannot provide the same cold, dry, bracing air of Northern

PREDICTION 1925 American scientists have discovered that the ultraviolet rays from the sun can cure children of rickets. Artificial sunshine lamps in clinics will help dull, stunted boys and girls put on weight, straighten up and become quick and intelligent in their movements.

Children still get rickets, especially in developing countries, and sunbathing—in natural or artificial sunlight—combined with vitamin D and calcium supplementation will cure them.

latitudes right at home, where the patients can breathe it continually, at temperatures suited to each case, while living in rooms flooded with sunshine.

PREDICTION 1924 **Prediction that a majority of surgical operations of the future will be performed while the patient is conscious** was recently announced before a gathering of doctors. By "blocking" the nerve group that controls the part of the body to be operated upon through the agency of a local anaesthetic, it was said that pain would be eliminated without the necessity of putting the patient to sleep. Considerable research work in nerve blocking has already been done, it was declared, and certain major operations have already been performed.

PREDICTION 1965 **Some researchers have gone beyond diagnosis and experimented with ultrasound as a treatment for various diseases.** Several cases have been reported where ultrasound loosened arthritic stiffness and broke up kidney and gall stones.

A daring and dramatic feat of this kind was performed in Italy, where doctors at the University of Padua reported a new concept in bloodless surgery. They described an operation in which three separate ultrasound beams were aimed at a deep-lying brain tumor. Like three invisible shafts the beams sliced through the

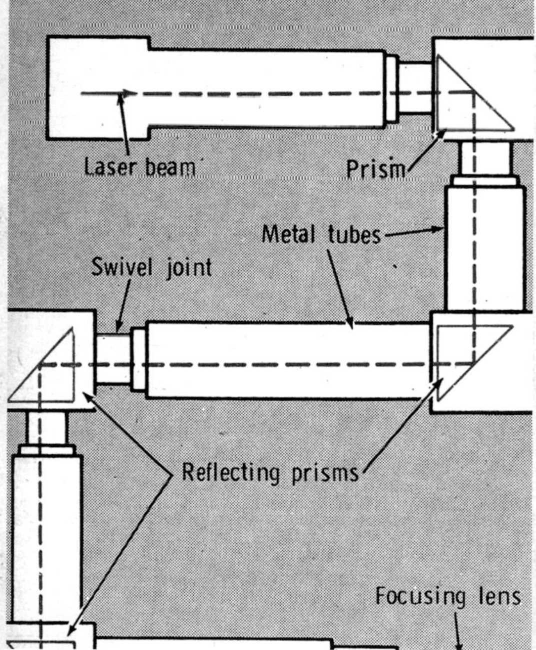

PREDICTION Looking like a dentist's drill, this swivel-jointed laser "knife" is used in surgery as if it were a scalpel. The beam is bent around corners by prisms so the arm can be moved in any direction. At the tip, it is focused to a needle-sharp point for cutting tissue.

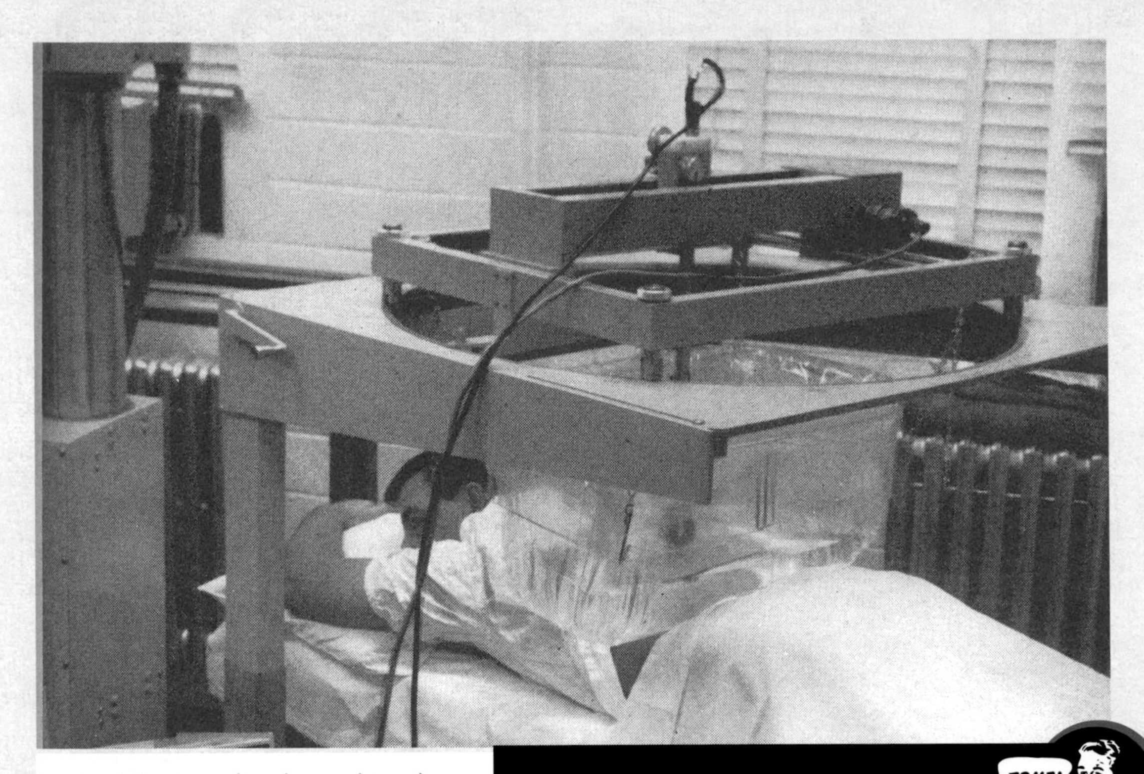

PREDICTION 1965 A traveling ultrasound transducer moves over the patient, enabling the doctor to get a cross section picture. Water provides fluid coupling for the sonic beam. By showing the condition of organs, the picture may eliminate the need for exploratory surgery.

TRUE!

Ultrasound today is used to diagnose a wide variety of medical problems. There are even storefront ultrasound operations that give parents-to-be keepsake images of their unborn children, though this practice has been discouraged by the FDA.

brain, but did no damage along the way. But when the three beams converged, the triple-strength ultrasound energy "jiggled" the tumor cells out of existence.

Research projects are now under way at the University of Illinois and Massachusetts General Hospital in Boston. The day may not be far off when some forms of surgery can be performed without incisions.

A CURE FOR THE COMMON COLD

PREDICTION 1925 Twenty-five years from now, tuberculosis will be as uncommon in the British dominions as leprosy is today, according to Dr. C.W. Saleeby of the London Medical Institute. He also predicts that infant mortality will be reduced to almost nothing and that successful ways will be discovered to combat cancer. Largely through preventive methods, the death rate from tuberculosis has steadily decreased in the United States. In 1910, the fatalities from this affliction of the lungs were nearly 140 for each 100,000 population and in 1922, less than one hundred.

PREDICTION 1938 **Removal of diseased organs from the human body,** their cure and then replanting in the body is the prediction of Dr. Alexis Carrel and Col. Charles A. Lindbergh. The work would be done with the aid of the Lindbergh pump, which bathes whole organisms from the animal body with life-maintaining liquids. Dr. Carrel, who helped to develop the pump, sees the day when the diseased portions of the body may be removed and sent to large Lindbergh pumps, as patients now are sent to the hospital. Hel believes replanting the organ in the body would offer no difficulty.

PREDICTION 1947 **Vein banks may some day take their place alongside blood and bone banks.** Dr. Charles A. Hufnagel of the Harvard Medical School reports that veins have been transplanted successfully in experiments with animals. He also reveals that a rapid-freezing technique makes it possible to preserve veins. He predicts the day will come when human veins are preserved and kept ready for use in persons suffering from vein injury or disease.

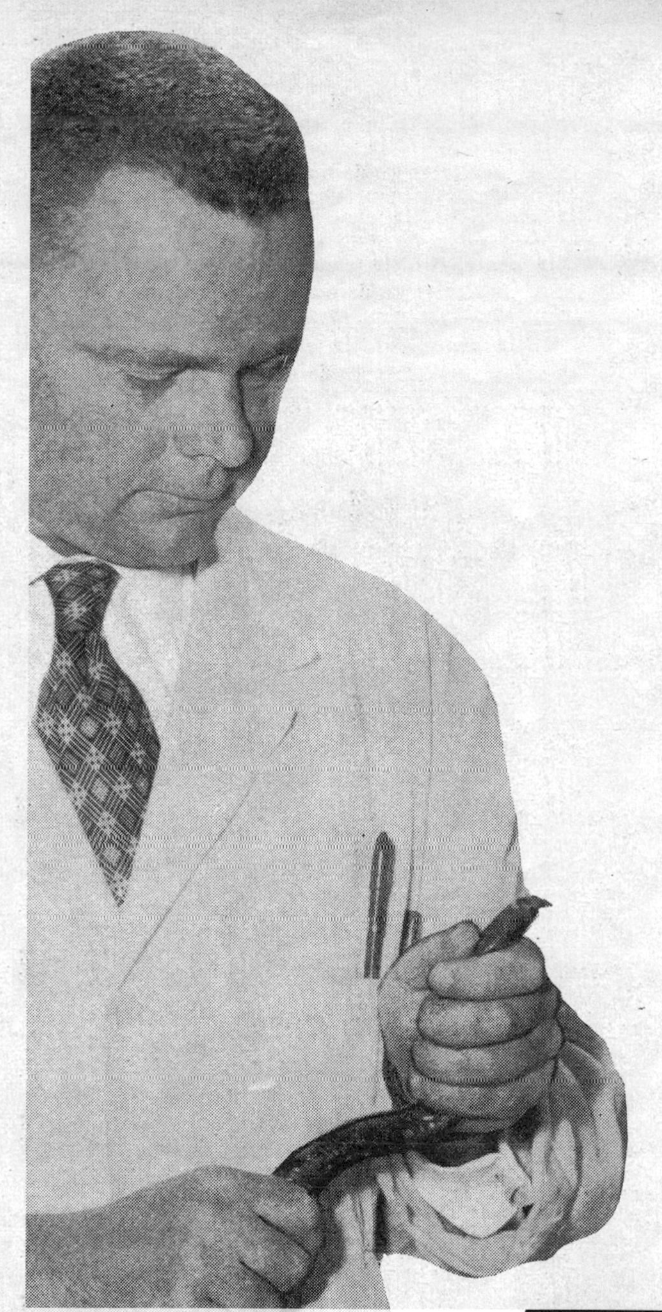

The Pacific hagfish has three hearts, one of which has no nerve connections to the body. Dr. David Jensen discovered that this nerveless heart is kept beating by a powerful biochemical pacemaker, eptatretin, which may replace implanted electronic pacemakers as a treatment to regularize the heartbeat of people with faulty cardiac nerves.

PREDICTION 1966

TOOTH TRANSPLANTS TO REPLACE FALSE TEETH

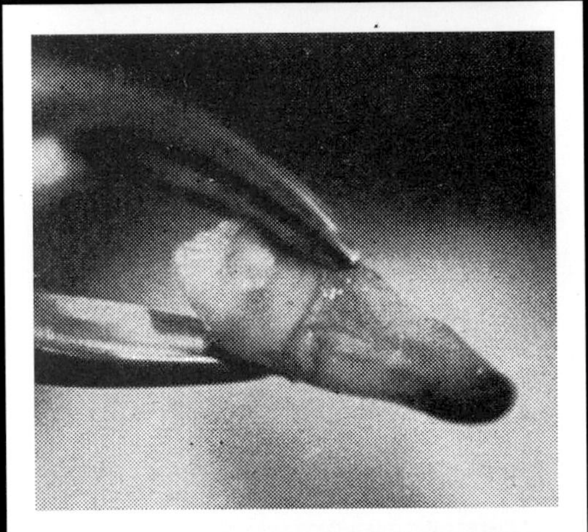

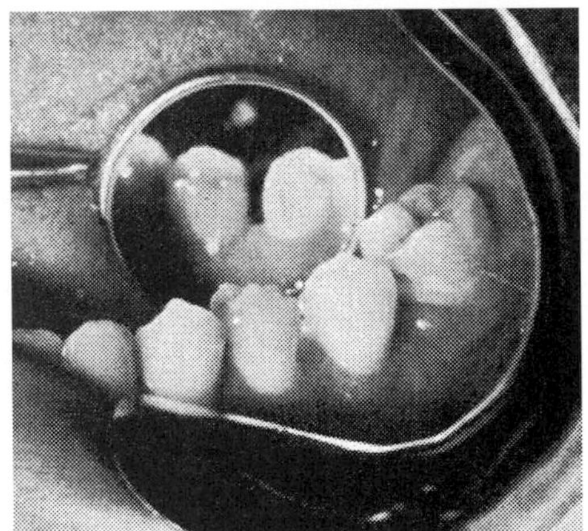

Extracted from a donor's overcrowded mouth, a tooth is quickly transplanted into the jaw of a second patient. Four weeks after the operation, the tooth (second from right) is fairly firm in its new socket and functions effectively in chewing. Teeth stored in tooth banks will one day replace false teeth. "Sooner or later," says Dr. S. Sigmund Stahl of New York University, "somebody will come through with an answer to transplanting kidneys, and when he does, we'll know how to transplant teeth."

PREDICTION 1950 In Year 2000, it is no longer necessary to administer the purified extracts of molds to cope with bacterial infections. Antibiotics are all synthesized in chemical factories and it is possible to modify their molecular structure so they acquire new and useful properties. By tying together what chemists have discovered about the structure of protein and what the pathologists see in the electron microscope, such virus diseases as influenza, the common cold, poliomyelitis and a dozen others are cured with ease.

While cancer is not yet curable in 2000, physicians optimistically predict the time is not far off when it will be. However, such afflictions as multiple sclerosis, palsy, or Parkinson's disease are no longer regarded as incurable. Sufferers from damaged or degenerate nerves must carry a little battery-driven apparatus in the pocket to provide the stimulus the nerves need.

PREDICTION 1959 Four young British scientists have isolated the basic penicillin molecule. Their discovery may lead to the production of countless "tailormade" varieties of penicillin, capable of knocking out organisms that are resistant to existing antibiotics. New varieties would also be valuable because many patients are sensitive or allergic to the antibiotics now available. The development of "thousands of new types" of penicillin would mean that a type could be found that all patients could tolerate.

Radiation can prevent potatoes from sprouting, and its anti-aging effects may soon be felt on people as well. Dr. Edgar J. Murphy, formerly chief of pilot-plant operations at Oak Ridge, recently pointed out: "It may be possible to find ways of using correct dosages and proper applications of radiations to extend the period of youth."

THE BREEDING OF HUMAN BEINGS

PREDICTION 1928 In a few generations, almost all persons will have brown eyes, a London specialist predicts. He bases his theory on the fact that brown eyes are better adapted to strong lights than blue, and that nature will therefore produce a brown-colored iris in the eyes of people who habitually face the strong illumination of artificial lights. The eye was not intended for the uses which civilization demands of it, the doctor holds, and, consequently, a natural change to fit the conditions will occur, or individuals will have to wear goggles with colored lenses.

PREDICTION 1932 To be sure, the scientific achievements of the next fifty years will be far greater, more rapid, and more surprising, than those we have experienced thus far. Startling developments lie already just beyond our fingertips in the breeding of human beings and the shaping of human nature. There seems little doubt that it will be possible to carry out the entire cycle which now leads to the birth of a child, in artificial surroundings. Interference with the mental development of such beings, expert suggestion and treatment in the earlier years, would produce beings specialized to thought or toil.

PREDICTION 1937 **Ten years or more may be added to man's life by drinking heavy water,** predicts Dr. James E. Kendall, head of the chemistry department at Edinburgh University, in Scotland. Heavy water, containing the heavy hydrogen atom, has the same effect on the body as lowering its temperature without actually doing so. It would slow functional processes, reducing bodily wear and tear without appreciably impairing man's faculties. Dr. Kendall believes persons over sixty soon will drink heavy water to slow the pace of life and to prolong it.

PREDICTION 1950 **Today's physicians do not know exactly how a piece of beefsteak is converted by the body into muscle and energy.** The physician of 2000 knows just what diet is best for a patient. This knowledge, coupled with his knowledge of hormones, enables him to treat old age as a degenerative disease. The span of life has been lengthened to 85, and men and women of 70 in A.D. 2000 look as if they were 40. Wrinkles, sagging cheeks, and leathery skins are curiosities or signs of neglect.

PREDICTION 1939 **By 1980 the average American may expect to live until he is 70 years old,** say statisticians of the Metropolitan Life Insurance company—and this mark will not be exceptional, but the average age reached by individuals. At the time of the Louisiana Purchase Exposition In 1904, life expectation was 48.2 years, and then it jumped to fifty-three years when the Panama-Pacific Exposition was held in 1915. The sixty-year mark was passed by the time of the Century of Progress in 1933 and 1934, and today, with two simultaneous World's Fairs as the milestone, it is estimated that an additional year has been gained, bringing the figure to 61.5 years for a white male and 65.2 for a white female.

TRUE!

American male life expectancy hit 70.1 years in 1980, exactly as predicted. For women it was an even more impressive 77.6.

This artificial placenta system for prematurely born lambs uses a new type of artificial lung for respiration. Perhaps some babies too will be "grown" during the last few months of their "intrauterine" life—in a "fish tank!"

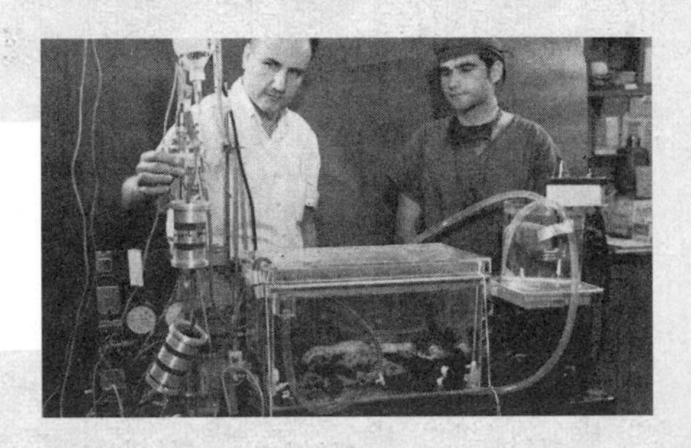

PLUG-IN PLASTIC ORGANS

PREDICTION 1937 By making the whole body an "ear," scientists are teaching the deaf to hear and interpret sound by its vibration patterns. It is predicted that within a few years doctors will know how to fasten a little vibration instrument to the toe of a child born deaf and enable the baby to overcome its handicap. Deaf children today are taught vibration patterns of speech and music by placing their fingers on a "phonotactor" into which the teacher speaks or music is played. The vibrations are carried through the skin and along the bones to the brain centers.

PREDICTION 1948 Electronic "sound-readers" for the blind may some day replace the Braille system. Under development at the University of Michigan, an electronic pencil converts the printed letters into sounds. These sounds vary for each letter and the blind person learns to identify them and thus to read. An improved model will actually read to the user, pronouncing the letters rather than making identifying sounds that must be memorized. The pencil, slightly larger than a fountain pen, contains a photoelectric cell that scans the printed lines. Sounds are amplified into a hearing-aid earphone.

PREDICTION 1966 If you're having trouble sleeping, relax and get a sleep machine for your bedside. Edward Ashpole reports from Britain that D.R. Garner & Co., Ltd., has developed and is manufacturing a sleep-inducing machine. The insomniac just wears two headbands, each containing silver mesh electrodes. The machine gener-

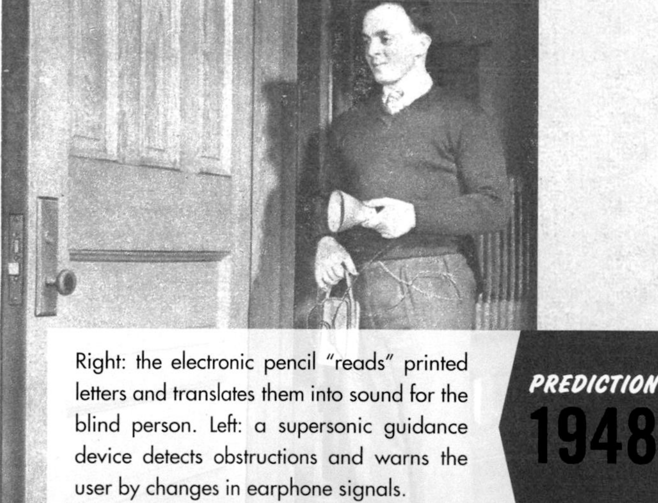

Right: the electronic pencil "reads" printed letters and translates them into sound for the blind person. Left: a supersonic guidance device detects obstructions and warns the user by changes in earphone signals.

PREDICTION 1948

ARTIFICIAL HEART AND LUNG MACHI

BLOOD DRIPS THROUGH MESH PACKING, ABSORBING OXYGEN AND GIVING OFF CARBON DIOXIDE

BLOOD FLOWS FROM BODY TO MACHINE

MAIN ARTERY

DISTRIBUTOR HEAD

HEART

MAIN VEINS

BY-PASS VALVE

BLOOD FLOWS BACK TO BODY

MESH PACKING

TUBE CARRIES REFRESHED BLOC BACK TO MAI ARTERY

FILTER

MECHANICAL HEA SIMULATES ACTIC OF NATURAL HEA

OXYGEN INLET

ARTIFICIAL LUNG

VANTON MECHANICAL-HEART PUMP

BLOOD OUTLET

PREDICTION 1952 This artificial heart and lung machine may someday help to keep your blood pumping during heart surgery.

ates square wave impulses which are variable in frequency and intensity.

The frequency needed to doze off varies with the individual; some people drop off with 20 pulses per second, others need 100. Brain recordings show that wave patterns of natural and electrically induced sleep are identical; this isn't the case with drug-induced sleep, which is often accompanied by side effects.

What'll they think of next for us of frail flesh and bone!

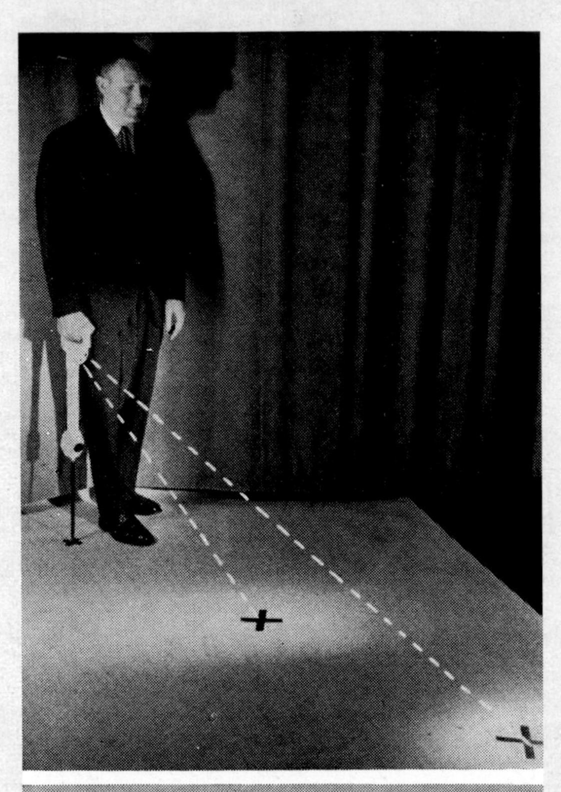

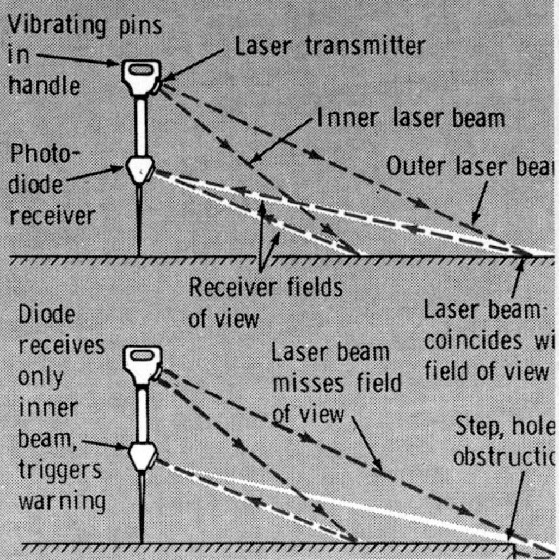

PREDICTION A "seeing-eye" cane spots obstructions for the blind. It sends out two laser beams three and six feet ahead. If beams bounce back unhindered, pins in the handle vibrate, indicating that the path is clear. If a beam is interrupted by an obstacle, the pins stop vibrating as a warning.

PREDICTION 1967 Dr. Adrian Kantrowitz, chief of surgery at Brooklyn's Maimonides Hospital, has developed a promising "heart booster," or auxiliary ventricle as he calls it. The device works by taking the load off the left side of the heart. The whole booster fits inside the chest; nothing sticks out except the power plug-in.

The idea is to let the patient go normally about his business. Whenever his natural heart needs a rest, the booster is ready for action. All he has to do is plug in the power supply into a plastic socket in his chest.

The Kantrowitz booster has no valves. Backflow is checked by the heart's own aortic valve and putting a tourniquet on the aorta in the bypassed stretch.

When artificial hearts are made on a production basis, the cost may be as low as $3000. The main problem is not hardware, but installation. "This surgery is far from a clinical routine," says Dr. Michael E. DeBakey.

Dr. Kantrowitz likes to look beyond the initial problems. "There is no technical reason why artificial hearts shouldn't be as commonplace as artificial legs."

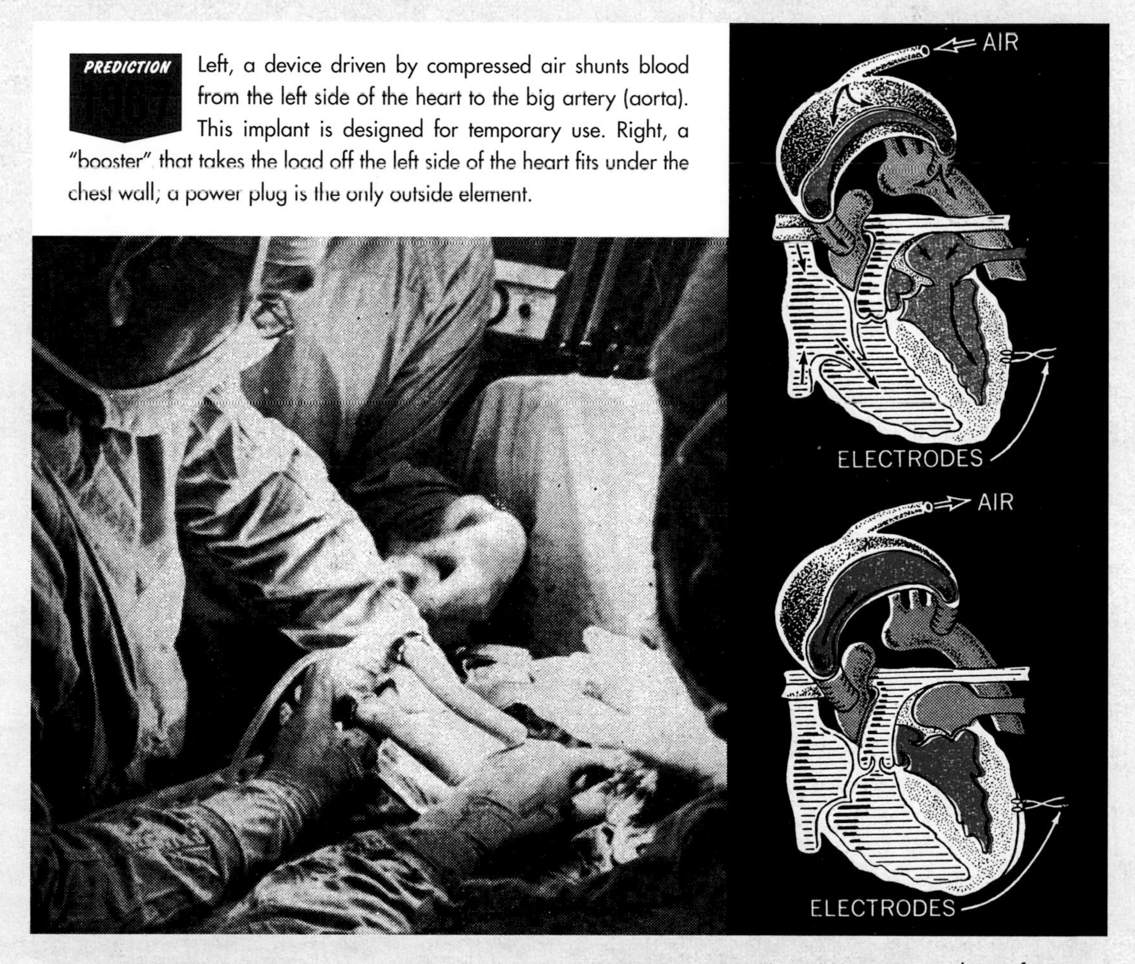

PREDICTION Left, a device driven by compressed air shunts blood from the left side of the heart to the big artery (aorta). This implant is designed for temporary use. Right, a "booster" that takes the load off the left side of the heart fits under the chest wall; a power plug is the only outside element.

5

AIRSHIPS SUPERSEDE BATTLESHIPS

The early twentieth century was "air minded." By 1920 everyone knew that flight was going to be the next really important technological revolution. Investors tried to find ways of making money from it, showing off their cutting-edge savvy. This resembled politicians of today, who will never pass up a photo op with a computer even if they don't even know how to type.

Airplane progress followed an S-curve. After a slow start there were rapid improvements in size and speed, so in about twenty years we went from biplanes to supersonic jets. Then speeds saturated, and passenger loads topped off at about five hundred. The limitations were in energy: going faster demands more energy, proportional to the square of velocity. To cut the flight time across a continent in half, you need to increase the energy expended by four. Chemical propulsion wouldn't send us any faster than the speeds we attained at midcentury. So airliners today are actually slower than they were twenty years ago. But many more fly than before, because of economies of scale. The "jet set" once referred to the privileged, but now includes most Americans.

Of course, optimism overshot in other ways. The "steamless, smokeless battleships of the future" did appear, but not the "jet-propelled ocean liner" or the "jet-powered automated trains that skim along on a cushion of air." Nor did we dig a tunnel across the continent so that we could "travel at a speed of 2,000 to 5,000 miles an hour in a vacuum tube," avoiding air friction.

The car of the future did indeed lead to "teardrop motorcars," though we don't travel at 150 miles an hour in them—again, energy constraints, plus safety issues. A 1950s prediction that in 2016 "cars will flit back and forth on cushions of air, the wheels retracting upon starting" doesn't seem likely. We call such devices hovercraft now, and noise alone prevents their use nearly everywhere. The promise of "hands-off driving on the automated highways of the future" may come true within a few more decades. A robot car drove itself 200 hundred miles a few years ago, across open desert.

But where's our "personal aerial vehicle"—the "aerocar" and the "helicab"—most practical for us in congested cities, especially for home-to-office commuters? It turns out that people cannot become good pilots without years of training (envision the LA freeways spread across the sky), so there will be no Jetsons-like future. Aerocars *were* developed, though. One aerocar model was still flying as recently as 2008. But none of them took off commercially. Instead, we spend more time commuting now than we did fifty years ago—most of it stuck in traffic.

Air hops around the city? That happened for a few decades, but helicopters kept falling out of the sky in the San

The airliner of the future might be anchored off shore on a special landing buoy in regions where construction of hangars is impractical.

PREDICTION
1951

This simple, practical, foolproof personal helicopter coupe is big enough to carry two people and small enough to land on your lawn. It has no carburetor to ice up, no ignition system to fall apart or misfire: instead, quiet, efficient ramjets keep the rotors moving, burning any kind of fuel from dime-a-gallon stove oil or kerosene up to aviation gasoline.

Francisco Bay Area and Los Angeles. Most spectacularly, on May 16, 1977, one tipped over on top of the Pan Am Building in New York City. I happened to be walking a block away and saw a spectacular sight: a chopper dangling over the building's edge. Passengers were boarding to fly to JFK Airport when the operating rotor blades cut them down as the landing gear collapsed. A blade fell into the street a block away from me. That killed the idea for society, and for me.

The 1951 conception of an atomic-powered rocket came true around 1970 in both the United States and the Soviet Union. But neither has flown in space because of treaties signed in the 1960s.

A 1940s vision that "[i]n 1959 we will build an elegant space station to accommodate about fifty men, then set out in perhaps two personnel ships and a cargo ship" didn't happen because a straight shot to the moon without orbital assembly was energetically and economically cheaper.

So, a box score: techno-prophets get the vast increase in speed and capacity of transport right. However, they didn't allow for social forces—concerns over safety, cost, and nuclear contamination—that could and did block credible technical feats. The dance between society and its engineers steps both backward and forward, a foxtrot of progress. ◎

PREDICTION
1960

This 380-foot monster of a submarine, shaped like a slim blimp and armed with 16 nuclear-tipped Polaris missiles, will have lethal fists cocked and ready to explode with knockout power from depths of oceans throughout the world.

Designed to eliminate accidents due to errors of judgment on the part of the pilot, a new type of tailless airplane is declared to make flying virtually fool-proof. It is said to be stable at practically any flying altitude, and in many positions instead of one, whether on a level course or pitched upward at an angle so steep that an ordinary airplane would go into a stall—the cause of a large portion of all air accidents.

PREDICTION

1926

THE STEAMLESS, SMOKELESS BATTLESHIPS OF THE FUTURE

PREDICTION 1906 There are some wise old sea dogs who declare that the extreme limit in mammoth ocean passengers has already been attained, if not overreached. They maintain that the danger in case of accident or panic is very greatly increased in the case of the extremely large craft. *The American Shipbuilder* says: "We might say that their great size tends to their structural weakness and early decay... there is a limit to safety in ocean steamers as on every other class of carriers."

With the ocean greyhound no restriction to weight exists, and while the mammoth is more difficult to get in and out of port, this consumes but a small part of her time. We predict the 1,000-foot steamer within the next five years.

PREDICTION 1909 The possibility of the adoption of internal combustion engines for the propulsion of battleships has been discussed for the past two years by naval experts on both sides of the Atlantic. Now, according to the English press, the internal combustion engine for such purposes has ceased to be only a possibility and has become a fact. England's experiments, it is claimed, have been so successful that a huge battleship, 540 feet long, 88 feet broad, and having a displacement of 21,000 tons,

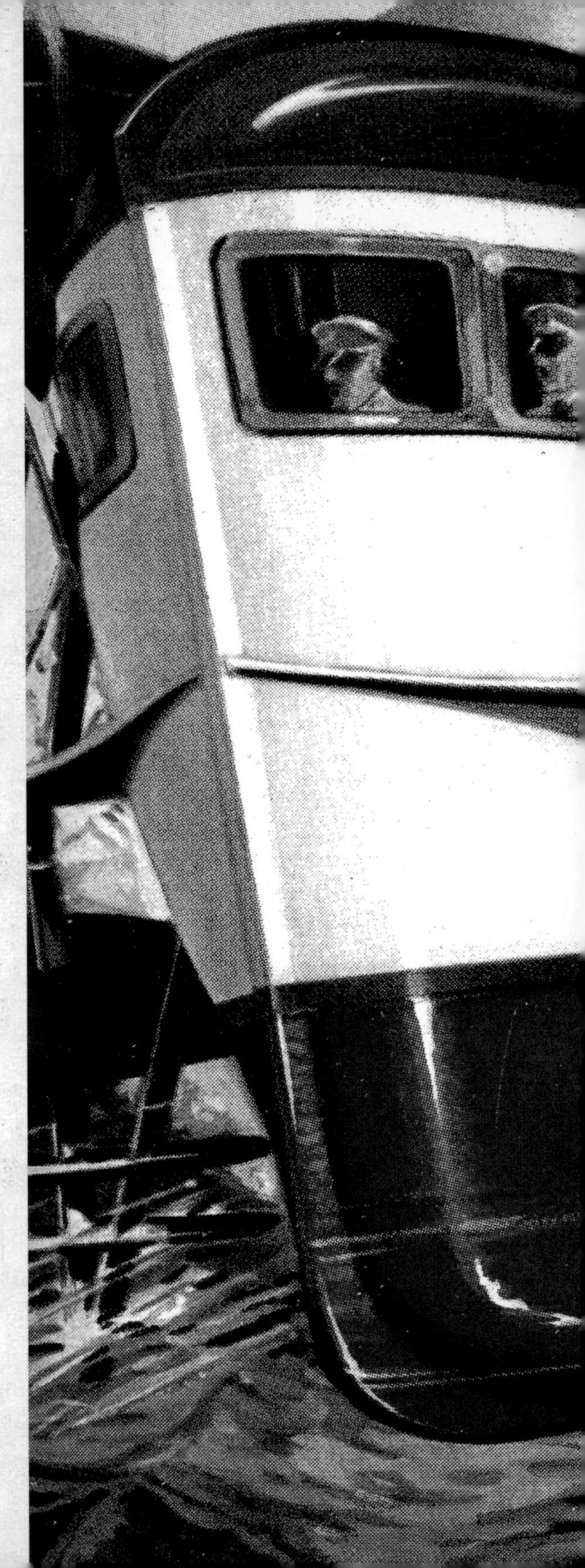

Accommodations for sixty passengers will be provided in a German glider boat driven by air propellers.
PREDICTION
1932

PREDICTION
1928

AMAZING ADVANCES IN *WATERCRAFT*

German Designers Have Planned This Streamlined Ocean Liner to Pull Itself through the Sea by Sucking in Water and Discharging It Farther Aft through Vents in the Hull

Left, This Queer Craft Is Not an Airplane, But a Hydroplane, Which Has Cut the Voyage from France to England to a Matter of a Few Moments; Below, a Proposed German Air Liner to Carry 200 Passengers across the Atlantic in Half the Time of the Fastest Steamers

PREDICTION
1946

JET PROPELLED OCEAN LINER OF THE FUTURE

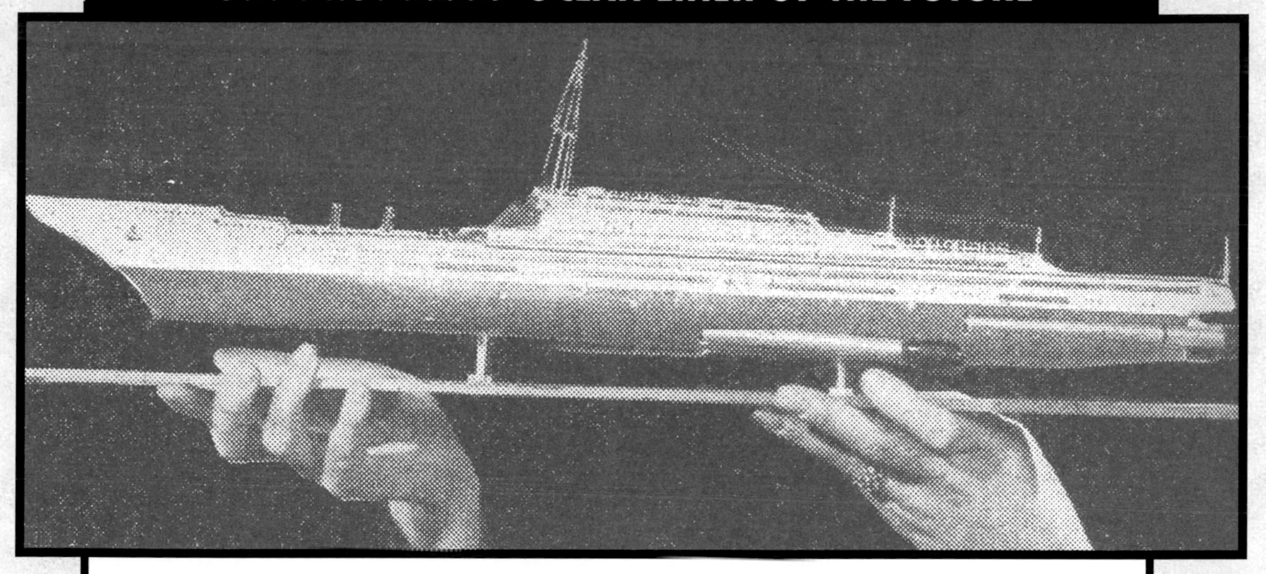

Jet engines will power passenger liners and merchant vessels of the future, according to designer J. Tomadelli. He has designed a scale model of a 1487-foot 10,000-passenger liner to be powered by four jet engines, two on each side of the hull. Estimated cost of such a 116,000-ton displacement vessel is $60,000,000.

TRUE!

Today's military ships use gas turbine engines, such as General Electric's LM2500, that are extremely similar to those used on commercial aircraft. These powerful engines consume high-quality fuel at a prodigious rate, making them too expensive for civilian use.

has been designed and will soon be laid down, if indeed the work has not already been secretly commenced.

The advantages of the internal combustion engine over steam power are numerous. There will be no smoke to draw the attention of the enemy, no funnels to obstruct the decks, the engines will be better protected, and oil tanks will be easier filled both at sea and in harbor than bunkers with coal. It is the absence of funnels, however, that will cause the most radical change in the exterior appearance of the war dogs. Except for masts, bridges, and conning towers the decks will be practically clear, making possible the arranging of big guns for firing in any direction. The huge funnels now in use offer fine targets to the enemy, their injury may reduce the draught and so the speed of the vessel, and their fall would probably put several guns out of action.

PREDICTION 1965 In water travel, the most significant advance since the steamboat has arrived, the hydrofoil liner which lifts from the water and rides on low-friction, airfoil-shaped "legs." Many hydrofoils are now operating in such varied locales as the Rhine River, on Italian and Swiss lakes, the Mediterranean, Baltic and Adriatic Seas, and are carrying passengers between Manhattan and the New York World's Fair. The U.S. Navy and Maritime Administration are experimenting with hydrofoils; one, built by Grumman Aircraft Corp., an 80-ton ocean going craft, and another, built by the Boeing Co., an awkward-appearing test vehicle designed for 115 knots!

JET-POWERED AUTOMATED TRAINS SKIM ALONG ON CUSHION OF AIR

PREDICTION 1905 A locomotive without water, fire or smoke, unencumbered by the five tons of coal and 7,000 gallons of water usually carried by the steam engine; drawing no tender, but provided instead with an engine for generating electricity; a clean, strong powerful engine drawing a 2,000-ton train and speeding across the continent from New York to San Francisco, without stop or delay, and at the average rate of 100 miles per hour—this is a dream

Monorails patterned after the Langen line in Germany, which will make possible speeds of 150 miles per hour between cities, and materially reduce traveling costs, will soon be in operation in New York.

The dream of modern traction—showing interior of locomotive.

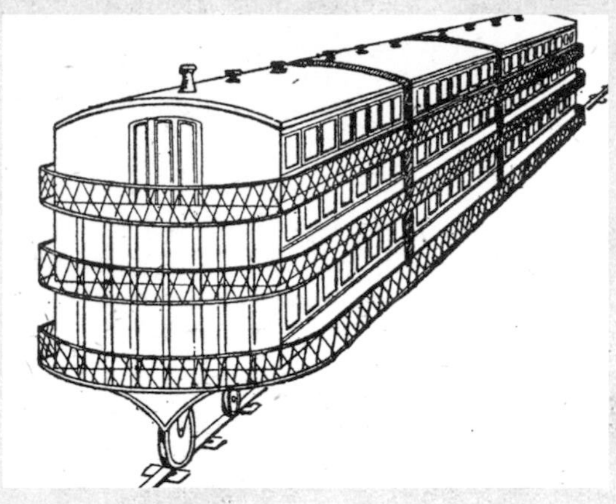

PREDICTION 1907

ABOVE: Monorail car of the future. BELOW: The car crosses a chasm on a single wire.

of modern traction, worked out theoretically with mathematical precision, now being constructed for the Southern Pacific Railroad and soon to be put to the practical test of a long trial run.

In the proposed new locomotive an internal combustion engine uses the crude oil that costs from three to five cents a gallon. The cost per horsepower hour is said to be less than half that for steam.

PREDICTION 1907

Louis Brennan, an English inventor, has astounded the scientific experts with his demonstration before the Royal Society. The inventor predicts the railway car of the future will be several times as wide as now, and two or three stories high. It will travel upon a single rail, and cross rivers on a single steel cable if conditions do not favor the use of piles or

PREDICTION

1907

GYROSCOPES WILL KEEP MONORAIL FROM TIPPING OVER

1. The active principle of the Brennan car: The Gyroscope in its simplest form, as it is sold for a toy—"The top that can't be knocked over." 2. The vehicle on the part of the track representing a mountainous district, showing the way in which it leans automatically towards the center in rounding a curve. 3. The vehicle on level ground, showing the way in which it leans automatically away from the heavier side when it is unevenly loaded. 4. The inventor at home. 5. A Blondin feat: The car crossing an iron hawser in the 6-ft. working model in Mr. Brennan's grounds, 5 ft. from the ground and keeping perfect balance. 6. The model car carrying a 150-lb. man. 7. The Brennan gyroscope. 8. Mr. Brennan's idea of travel in the future: A monorail vehicle very much larger and wider than present-day railway carriages. 9. Detail of the car.

piers. The propelling power may be steam, electricity or gasoline. If Brennan's expectations are realized his system will revolutionize the operation of railways throughout the world.

The secret of the Brennan system is the use of a gyroscope within each car. He has studied this mysterious piece of mechanism for 30 years and is said to be one of only three men in the world who really understand it. He says:

"Each vehicle is capable of maintaining its balance upon an ordinary rail laid upon ties on the ground, whether it be standing still or moving in either direction at any rate of speed, notwithstanding the center of gravity is several feet above the rail and the wind pressure, a shifting load, centrifugal action, or any combination of these forces may tend to upset it.

"Everything points to a great economy resulting from making the cars wider in proportion to their length than on ordinary railways. Therefore it has been decided to make an experimental coach 12 ft. wide.

"The speed can be from twice to thrice that of ordinary railways, owing to the smoothness in running and the total absence of lateral oscillation."

Until the government tests Brennan's device, this subject will continue to be one of absorbing interest to engineers and scientists, some of whom pronounce the invention the "greatest since the electric motor."

PREDICTION 1928 A new campaign in the conquest of time and space is to be launched in Germany within the next few months, when work starts on the first section of a 200-mile-an-hour monorail line which eventually will link Berlin with the industrial district of the Ruhr valley. Streamlined torpedo-shaped cars, driven by airship propellers at either end, are to be hung on ball-bearing rollers from an overhead track.

Monorail lines are not new in Germany, but the novelty of this new project is the great speed expected from the use of air propellers instead of the usual geared driving wheels. Electric motors will turn the propellers, getting their energy from a trolley rail, so that no fuel will be carried. The aerial cars have been designed by engineers of the Schuette-Lanz airship company, one of Count Zeppelin's former rivals in dirigible building. The projected speed—200 miles an hour—is practically double that of passenger-carrying airplanes, nearly three times the most efficient cruising speed of Zeppelin airships, and almost four times as fast as the railway schedule for the same distance.

PREDICTION 1950 Imagine a tunnel with one end beneath New York City's Times Square. You enter a car at this end, stow your suitcase in the rack overhead and settle down comfortably with a magazine. You have been reading scarcely an hour when the vehicle stops. An escalator carries you

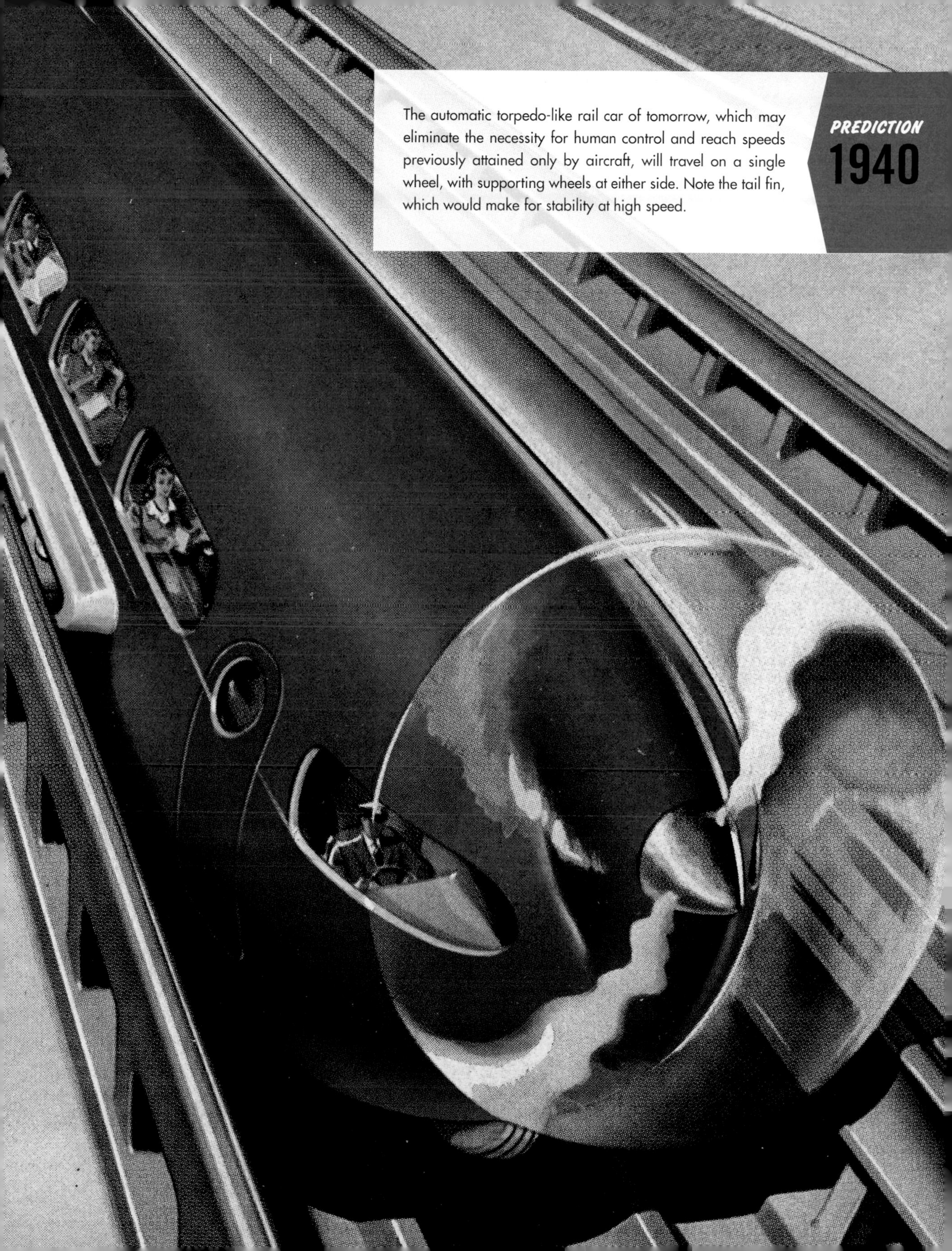

The automatic torpedo-like rail car of tomorrow, which may eliminate the necessity for human control and reach speeds previously attained only by aircraft, will travel on a single wheel, with supporting wheels at either side. Note the tail fin, which would make for stability at high speed.

PREDICTION
1940

back to the street level and you greet the light of day once more—in San Francisco!

Sounds like something out of pseudo-science fiction, doesn't it? Yet it's the idea of one of America's most practical scientist-executives, General Electric's noted physicist, Dr. Irving Langmuir. "There is no fundamental reason," says Doctor Langmuir, "why we could not travel at a speed of 2000 to 5000 miles an hour in a vacuum tube. The Pacific coast might be only an hour away from the Atlantic."

PREDICTION 1965 Within our lifetime, we may see jet-powered automated trains that will skim along on a cushion of air at 200 mph and passenger-carrying capsules that will rocket from coast to coast through tubes at 2000 mph.

A British inventor, Christopher Cockerell, has built a six-ton model "Hover-train" with a 240-hp jet engine to blast it into a hovering position a half-inch from vertical and horizontal surfaces of a T-shaped concrete rail. From that position, even a child's push can send it skimming

PREDICTION This proposal for a "tube train" goes back to the old pneumatic-tube system used years ago in department stores. A capsule is pushed into a vacuum tube and the air pressure behind it literally "whooshes" it to the other end. In the tube train, which runs on rails, the en route portion of the tube is evacuated of air while the station areas are at normal atmospheric pressure. A valve is opened allowing the train to enter the tube, and the air pressure behind it accelerates it up to 300 mph. The valve allows just the right amount of air to push the train to its next station.

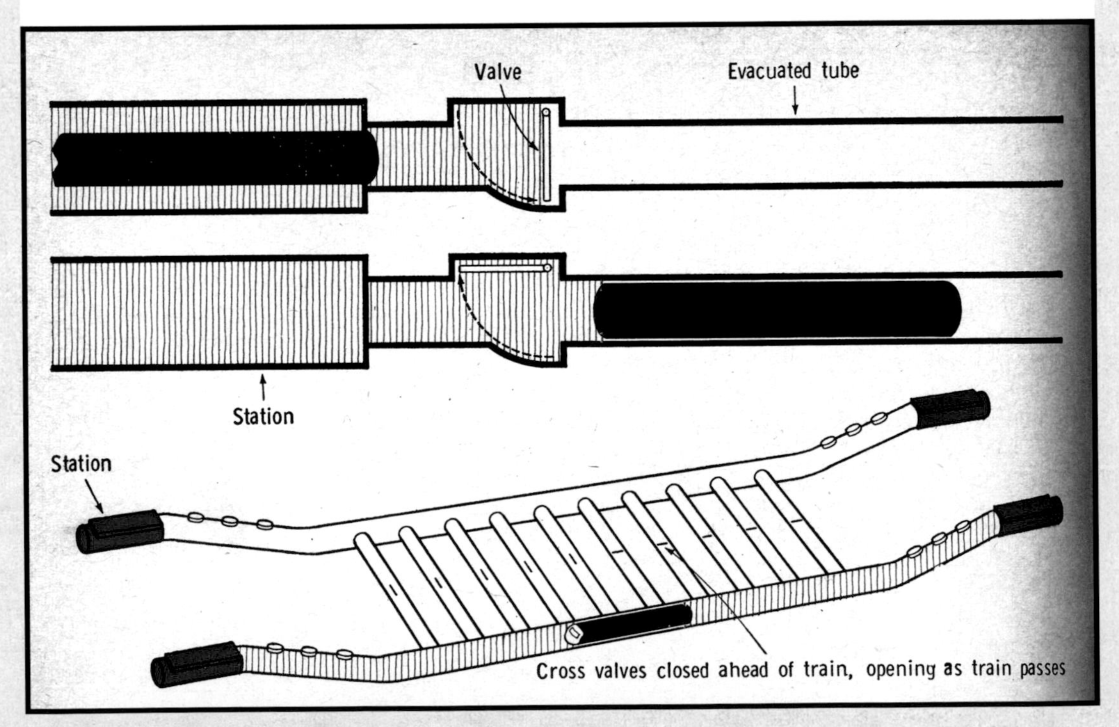

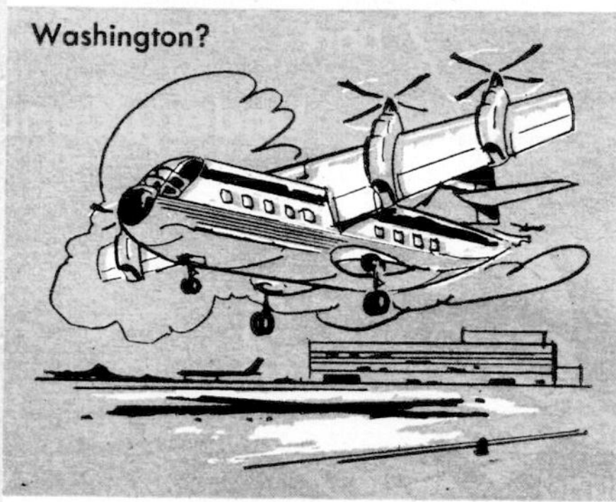

PREDICTION New systems for getting to and from outlying airports, and avoiding bottlenecked street traffic, include (from top) monorail trains, which would carry large numbers on direct trips; STOL and VTOL aircraft, which fly above traffic; rubber-tired trains that operate automatically without motormen, and combination bus-trains that ride either on tires or rails.

frictionlessly over the rail. With jet power or a linear induction motor, which has no moving parts, such a train easily could travel 400 mph, he claims.

Even more revolutionary is a longdistance transportation tube projectile system designed by Dr. Joseph Foa, head of the department of aeronautical engineering and astronautics at Rensselaer Polytechnic Institute. "Land vehicles are necessarily constrained to travel along narrowly defined paths," he says. "Therefore, unless the prescribed path is a straight line, a land vehicle traveling at high speed must be supported from all sides. This kind of support is best provided within a tube."

PREDICTION 1968 Communities are studying new and novel ways to carry passengers out to the airport. Some airports hope to reduce, if not eliminate, auto traffic by building more attractive rapid-transit systems. Mass transport of hundreds of passengers in just one vehicle, instead of in scores of automobiles or taxis, is an appealing proposition.

Tokyo already has a monorail to the airport, and other cities are considering such a system. Paris, which is now building its new Paris-Nord Airport far out in the country, plans a direct connection with its subway system. New York has also examined the feasibility of extending its subway system out to its bottlenecked Kennedy and LaGuardia Airports.

TRUE! *The A subway line now extends all the way from the northern tip of Manhattan to Kennedy Airport, where passengers can transfer to a monorail-like train to visit the various terminals.*

Because one of the major obstacles of attaining high speeds in other type vehicles—drag from air made turbulent by the act of ramming through it—would be eliminated, great speed could be attained with remarkable little power.

This system has been tested in model form. Now Dr. Foa hopes to build a pilot version in a tube three miles long.

THE MOTOR CAR OF THE FUTURE

PREDICTION 1903 The automobile review, discussing transcontinental automobile trips, says: "The great feat of crossing the continent in an automobile, will in a short time be a thing of the past and the trip is destined to become a summer outing for the enthusiastic automobilists."

It is likely, however, that a person will not be over-anxious for more than one trip in a lifetime.

PREDICTION 1940 The body of the car of the future, even to its windshields and windows, will be of new synthetic materials, probably some forms of molded plastics. In this car will be found the last word in devices that will make for the safety of driver and passengers—windows and windshields of clear, transparent substances, which neither shatter nor fly under impact and thus do not cut, and which also

Outstandingly different will be the motor car of the future—a future that is not far distant. It will carry its engine in the rear, where the engine has belonged all along, and will offer entirely new concepts of comfort—"living room luxury"—with new spaciousness, radically changed seating arrangements, complete summer and winter air-conditioning system, and many other innovations.
PREDICTION
1940

PREDICTION

1914

A novel vehicle has reached speeds of 67 miles per hour upon the boulevards of St. Louis. Within a huge wheel or ring of aluminum bearing a solid rubber tire is a curious framework of steel tubing that remains stationary while the wheel revolves. The air propeller, nearly 5 ft. long, is driven by a three-cylindered engine to provide forward motion. A small wheel or roller gives the machine stability until enough momentum has gained to enable the rider, by inclining his body forward, to lift the roller clear of the ground. There is no steering device, the inclination of the body of the rider being sufficient, as in bicycle riding.

admit the healthful ultraviolet rays of sunlight and exclude the infrared; airplane-type "crash pads" designed right into the interior to protect occupants from hard bumps in road accidents; a strong, spring-mounted bumper, running entirely around the body; and an integral and pleasing part of the over-all design, capable of absorbing far more of the shattering force of collisions than bumpers do now.

TRUE! ***Automobile glass today is designed to come apart into safe, jigsaw-puzzle-like pieces, and other tremendous strides have been made in the engineering of automobile bodies that will keep passengers safe in an accident.***

PREDICTION 1965 Out on the highway, a new era is about to embark: the automated trip. General Motors' proposed "Autoline" is a complete speed and directional control system for vehicles, using remote-control electronic signals. Its main components are the inevitable computer, a control cable beneath the automated highway lane, and sensing devices and servo-mechanisms in automobiles that can actuate their controls. When perfected, Autoline will let drivers "phase" into an automated expressway lane with their cars completely in command of the auto-line control. Cars thus can speed along, virtually bumper to bumper, at 50 mph.

STOP AND GO LIGHTS ON DASH TO *CONTROL AUTOS*

Driving through the crowded streets of the future, autoists will stop, wait for cross traffic to clear, then start again without the aid of today's signal post on the corner. Stop-and-go light signals, installed on the automobile dash, will control tomorrow's traffic. A glance at his instruments will show the driver whether the way is open long before he reaches the next street intersection.

The car of the future, by virtue of its greater roominess, can be made literally a luxurious "home on wheels." Seats can be converted into comfortable berths for sleeping. In a twinkling, a portable table can be set up. There will be plenty of convenient storage places, even little refrigerated cupboards for foods and drinks.

PREDICTION 1956

ITALIAN DESIGNERS PLAN 125-MPH *BUS OF TOMORROW*

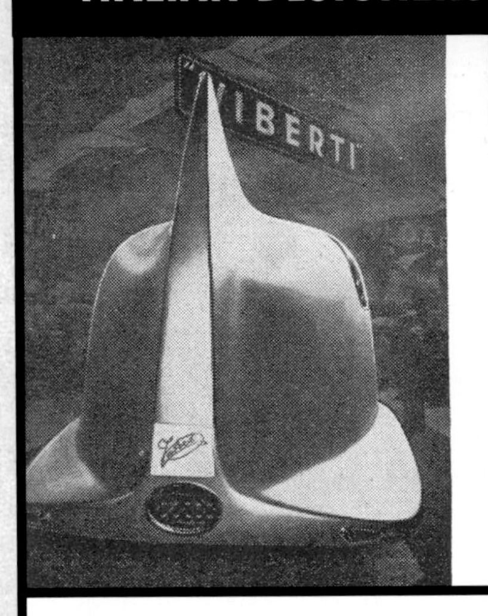

Extreme lightness and a 125-mile-per-hour speed are claimed for an ultramodern bus designed by the Viberti Company of Italy. The bus will eventually be powered by a gas-turbine engine, but a test model is being made with a conventional piston engine. To achieve lightness, the vehicle is built of plastics without any metal framing. Independent suspension is featured on all four wheels. The tail is like that of an airplane. The interior is air conditioned and has facilities for serving food and watching television. The driver's seat is exactly in the center for good visibility.

PREDICTION 1966

What will cars be like in 2016? The Automobile Club of Michigan, this year celebrating its 50th anniversary, wondered about this recently and approached widely publicized seer Jeane Dixon, asked her to gaze into her crystal ball and come up with a few answers. Fifty years from now, Miss Dixon predicted, cars will flit back and forth on cushions of air, the wheels retracting upon starting. They will be fueled by some exotic new compound yet to be developed; gasoline as we know it will have gone the way of the buggy whip. A radar-like device will guard against cars being involved in accidents. Consensus here is that Miss Dixon is on fairly safe ground as "studies" of such designs and gadgetry are already in the works, although a long, long way from fruition.

PREDICTION 1967

Acceleration, braking and steering are all combined in this single airplane-type control—a first step toward hands-off driving on the automated highways of the future.

Push forward and you accelerate; pull back and you brake; lean it to the left or right and off you go in that direction.

PREDICTION
1940

Above is one conception of the car of the future, drawn especially for ***Popular Mechanics***. Note its circular lounge in the rear and the wide visibility offered by its transparent plastic roof.

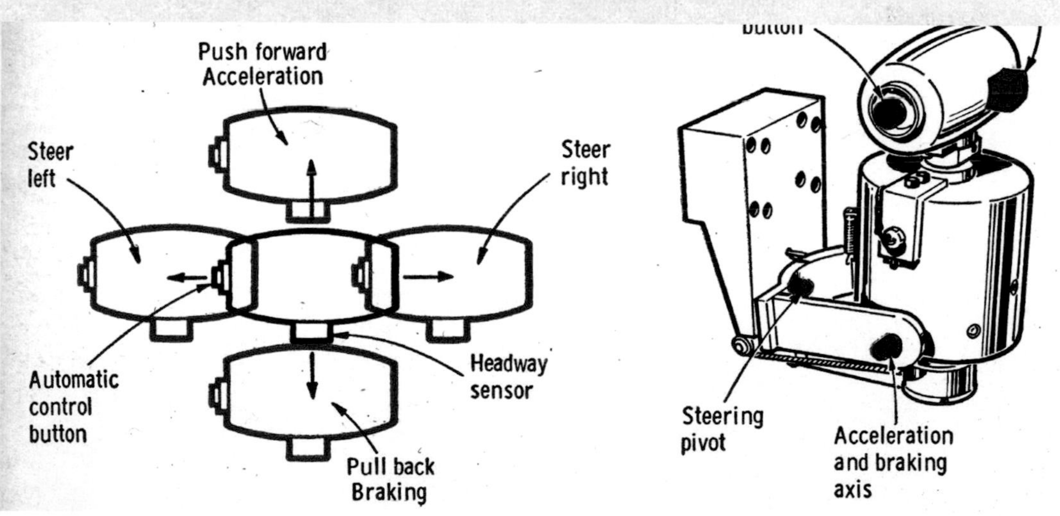

PREDICTION Stick operation and configuration prototype design. Further development of automatic highways will apply human factors to design, operation and placement in future cars.

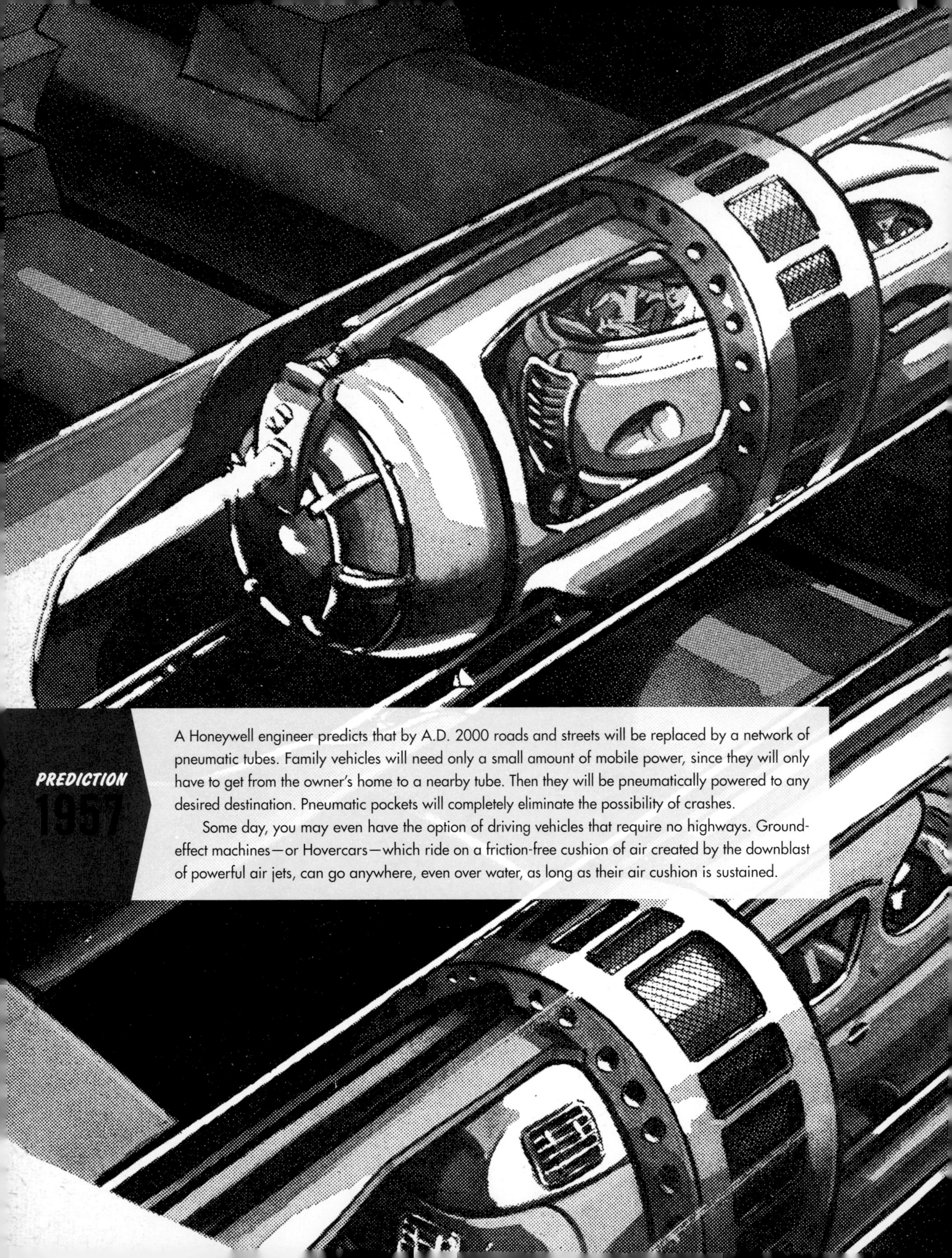

PREDICTION

1957

A Honeywell engineer predicts that by A.D. 2000 roads and streets will be replaced by a network of pneumatic tubes. Family vehicles will need only a small amount of mobile power, since they will only have to get from the owner's home to a nearby tube. Then they will be pneumatically powered to any desired destination. Pneumatic pockets will completely eliminate the possibility of crashes.

Some day, you may even have the option of driving vehicles that require no highways. Ground-effect machines—or Hovercars—which ride on a friction-free cushion of air created by the downblast of powerful air jets, can go anywhere, even over water, as long as their air cushion is sustained.

Among other gadgets, the car includes a "tactile headway sensor." Its purpose is to assist a driver in maintaining an accurate headway by adding the sense of touch. Before arriving at the idea of the tactile sensor, the laboratory investigated auditory and visual methods of informing the driver that he is drifting, but none was as effective as the tactile system.

The reason why? To accommodate growing traffic, cars on future highways will in tight. Cars will slow down, speed up, and, if lucky, drivers will reach their destinations after a jerky and fatiguing trip.

PREDICTION 1967 Mapless driving? Yep, and here's how it would work: At the start of a trip a driver hops in his car and dials a code number representing his destination. The number is read by route guidance equipment inside the car. The equipment automatically transmits the code to key roadside equipment, which then transmits—either by voice or visual display in the car—routing instructions to the driver.

TRUE! ***Rather than relying on roadside transmitters, today's GPS determines the car's location by triangulating with transmitters in satellites. Directions by voice or visual display are beginning to supplant the paper map.***

PREDICTION
1957

The "flying fan" vehicles of the future will be easier to fly than helicopters, and should cost a lot less. A rough guess is that in about 10 years you'll be able to buy a four-place fan for the price of a good car.

YOUR PERSONAL AERIAL VEHICLE

PREDICTION **1928** The airplane-automobile in the garage of the home of tomorrow has been carefully built by an airplane designer. The propeller, mounted in the nose of a streamlined, glass-enclosed car body, is mounted on a movable shaft which could be elevated ninety degrees, to place the propeller in a horizontal position above the car—for straight ascent as a helicopter.

PREDICTION **1943** Postwar transportation, according to W.B. Stout, will rely largely on the aerocar, or flying automobile for families; the roadable airplane, for distance flight accompanied by short trips on the ground, and the helicab, a new type of helicopter. Stout, renowned engineer and designer, is now head of Stout Research Division of Consolidated Vultee. He believes the public is ready for such machines, and that industry is ready to provide them.

His aerocar is a three-passenger family model which takes wing for week-end or vacation trips. Weighing 1,500 pounds—half as much as a pre-war light coupe—it will be a good automobile first, and a plane only second, capable of speeds of 70 miles an hour on the highway and 100 in the air with pre-war gasoline consumption. Its cruising range would be 250 miles. To convert it to a plane, you let down wing and outrigger assembly, hook it on, and fly away.

PREDICTION 1943 Unlike the roadable plane, which is intended to be primarily a plane and secondarily a car, the Aerocar, the flying family auto, is, first, a good automobile.

His roadable plane will be a good airplane first, and a ground vehicle second. Light, weighing only 800 pounds, it would be capable of 400-mile cross-country hops, and could be used as a light delivery truck—say in areas of the West, where communities are far apart. It will do 120 miles an hour in the air, but only about 35 on the ground. To convert it from plane to car, you land it on the road, fold up the wings, and drive on.

These two models would be primarily for the Midwest and West; in the more congested Northeast, and indeed in any metropolitan area, Stout's "helicab" is designed to provide the nearest thing to personal wings yet devised. A helicopter type, designed to carry up to five persons, it would weigh about 1,700 pounds. He foresees a half-million helicopters taking commuters to Los Angeles and New York each morning, and back to their suburban homes each evening.

Stout's Aerocar would do 60 to 70 miles on the highway, and 100 miles an hour in the air with about the same gasoline consumption. Its air range would be about 250 miles. His roadable plane would be for use in the West, where distances are longer, and would have a range of 400 miles in the air, coming to the ground only if the weather is bad, and remaining on the road only until the storm area is passed. Its wings would fold for road driving. It would speed at 120 m.p.h. in the air, 35 m.p.h. on he ground. Above is a third of Stout's conceptions for postwar travel, a Helicab. He considers it most practical for use in congested cities, especially for home-to-office commuters. Stout says he foresees electronic controls to prevent collision and constant ground-air communication

PREDICTION 1950 Instead of driving to California in their car—tear-drop in shape and driven from the rear by a high-compression

Inventor W.B. Stout considers the Helicab most practical for us in congested cities, especially for home-to-office commuters. Stout says he foresees electronic controls to prevent collision and constant ground–air communication.

engine that burns cheap denatured alcohol—the family of A.D. 2000 use the family helicopter, which is kept on the roof. The car is used chiefly for shopping and for journeys of not more than 20 miles.

The railways are just as necessary in 2000 as they are in 1950. They haul chiefly freight too heavy or too bulky for air cargo carriers. Passenger travel by rail is a mere trickle. Even commuters go to the city, a hundred miles away, in huge aerial busses that hold 200 passengers. Hundreds of thousands make such journeys twice a day in their own helicopters.

PREDICTION 1957 A new kind of flying machine is being designed that sounds like the answer to your desires for a personal aerial vehicle. It is almost like a flying carpet. A good name for it might be the "flying fan." It uses the principle of the ducted fan, the same idea that is used by the flying platform.

Superficially this new machine, now being developed by Hiller Helicopters, will resemble an automobile although it will rest on short stilts. You'll be able to order a four-door model, a sports job, or even a light-truck configuration.

PREDICTION

The Rotavion may be the world's safest aircraft. The spare-time project of several dozen airplane engineers and aircraft shop men, this vertical-takeoff-and-landing craft can cruise at 175 mph. in level flight, yet it will stay out of trouble even with a dead engine. The louvers above and below the ducted fan can close to convert the craft into an oddly shaped but efficient airfoil or open to permit helicopter-like operations.

PREDICTION 1944

HELICOPTER "AIRBUS OF THE FUTURE"

The Greyhound Corporation showed this model of a 14-passenger helicopter as the "air bus of the future." The model, which has a cigar-nosed, streamlined fuselage and a triangle landing gear, was shown recently at hearings before the Civil Aeronautics Board.

There's no big thrashing rotor overhead, no churning propellers. You ascend with no apparent effort. Air speed is up to 50 miles per hour. If you want to stop in mid-air, just move the lever back again.

All this sounds too easy to be true, but the designers say this description is about right. Probably the first flying fan will be in the form of an aerial Jeep, an aircraft in which the Army is greatly interested. Hiller Helicopters has submitted a proposal for a four-fan Jeep. Performance and costs have been closely estimated.

This Jeep-of-the-air would be capable of every job that an ordinary Jeep can do, with the advantage that it could travel across country with no need for roads. It would be able to land and take off from fairly steep hillsides. It could carry men and supplies, be used as an ambulance, or serve as a flying gun platform.

For this kind of service the flying fan has several major advantages over an ordinary helicopter. It can operate close to buildings or other obstructions. A helicopter taking off from alongside a building gets a rebound of air from the building that makes it very hard to control. And the flying fan can be operated around people in safety, with no chance that someone might accidentally walk into the whirling overhead rotor.

The Hiller engineers expect that eventually the ducted fan will become the basis for a whole family of special-purpose aircraft.

FLYING-MACHINES OF TOMORROW

PREDICTION 1909 "In less than ten years an aeroplane will cost no more than $500," is the prediction made by Frank Hedges Butler, the English balloonist, just returned from an ascent in Wilburt Wright's aeroplane. His glimpse into the future:

"Lighthouses on land will be erected by the Trinity board to mark the way at night. Lamps on aeroplanes or flyers will be used. With the smaller planes the speed will be terrific—200 miles an hour—and the 21-mile trip across the channel will take only a few minutes."

PREDICTION 1922 Consider yourself aboard a giant airplane whose whirring propellers rapidly drive from view faint objects on the earth far below. As towns and hamlets recede in the distance you realize that you are fast approaching the one that is your destination. The crew unfold from the capacious hold a small air boat, and lower it dangling from the huge hull by its special tackle. You and several fellow passengers climb down into the seats behind the pilot and buckle yourselves in as the big ship slows its engines to enable the little wings to catch the air. With a quick movement of a lever your steersman unleashes the small craft, which begins its motorless flight and gracefully glides downward to a safe landing, while the mother plane speeds out of sight.

Just as we have seen Jules Verne's imaginary voyage in "1,000 Leagues under the Sea" so nearly realized in the exploits of the submarine of today, it is not too far a stretch to consider air travel as a possibility. Even now, large high-powered airplanes of all-metal construction are being built under contract to carry as many as 100 passengers for 1,000 miles or more without stopping.

PREDICTION 1951 More money and fresh brains have been thrown into the top-secret U.S. project for harnessing atomic power for the aircraft and the target date for such a plane—bigger than a B-36—has been advanced (though it's still secret).

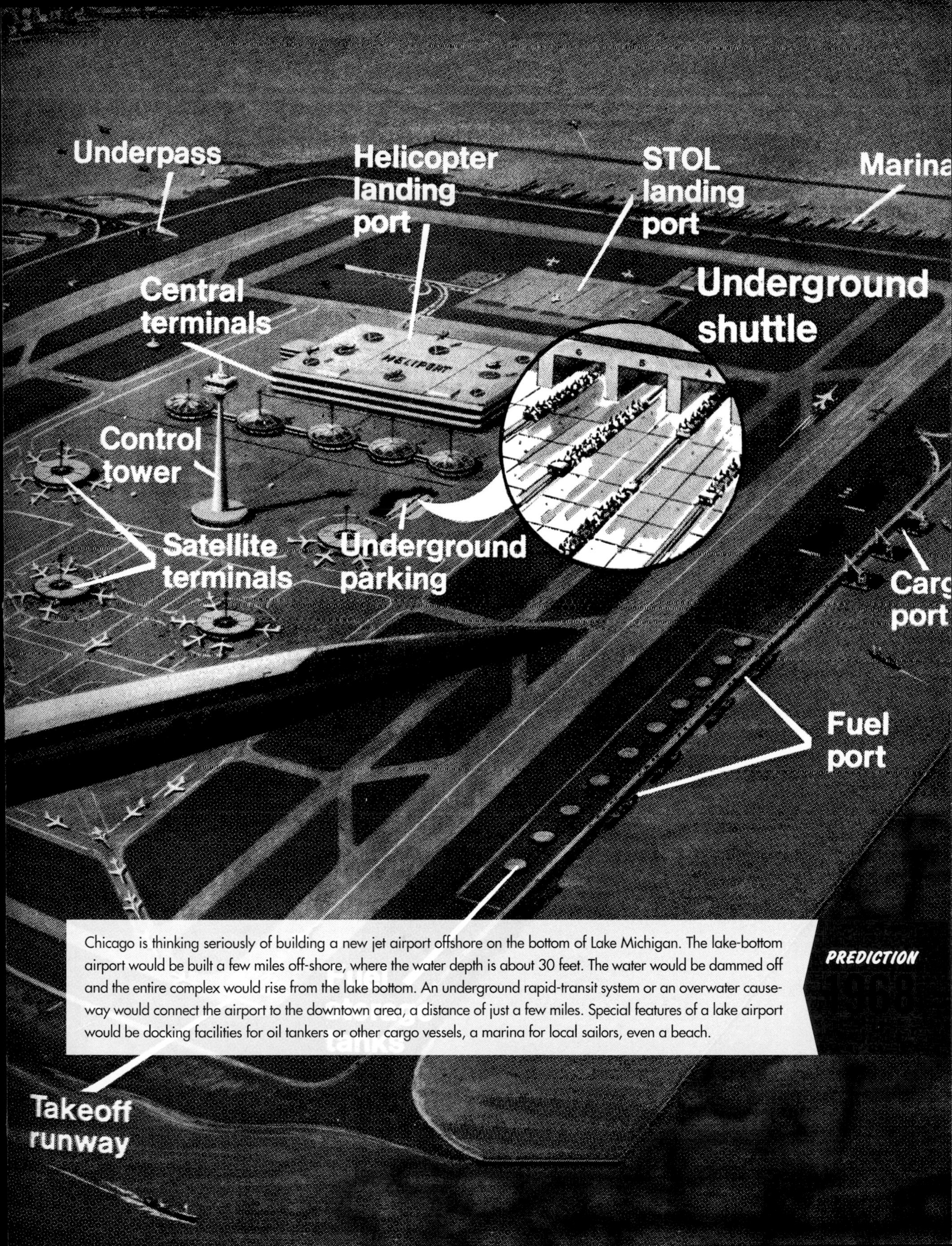

PREDICTION

Chicago is thinking seriously of building a new jet airport offshore on the bottom of Lake Michigan. The lake-bottom airport would be built a few miles off-shore, where the water depth is about 30 feet. The water would be dammed off and the entire complex would rise from the lake bottom. An underground rapid-transit system or an overwater causeway would connect the airport to the downtown area, a distance of just a few miles. Special features of a lake airport would be docking facilities for oil tankers or other cargo vessels, a marina for local sailors, even a beach.

PREDICTION
1907

HEAVY MOTOR-DRIVEN FLYERS

The motor-driven flyer is a success, but so far its relatively great weight has made experimenting dangerous because of the possibility of the motor giving out, or some accident happening to the steering or supporting surfaces. The Wrights' and Dumont's machines both had sufficient surface to make safe landings after accidents, but Santos' new machine, built entirely of aluminum and consequently very heavy, will be absolutely unmanageable if its motor stops. It is such machines as this that will shatter the public's faith in our ability to ever make flying safe. These projectile machines, depending entirely on their high-power motors for propulsion, will do very well for speed maniacs who are longing for new sensations and accompanying dangers, but most people will wait for a machine that will carry them through the air at a reasonable speed and without turning air-springs every time the motor feels indisposed.

ABOVE: Giant Transatlantic Flying Boat Proposed by a German Inventor. BELOW: Twelve-Engined Flying Boat Being Built in Switzerland by the Famous Dornier Factory.

PREDICTION
1922

Small motorless glider taking leave of giant mother airplane, to land passengers at an intermediate point along route of aerial highway, as proposed in Russo-German plan.

Even more intriguing than the atomic submarine, but also more complicated, is the idea of an airplane with a nuclear-driven power plant. It could fly nonstop around the world at supersonic speeds, ride out any bad-weather traffic stacked over an airport and would never be subject to a power failure on take-off or landing.

Materials-testing piles already have turned up effective shieldings of very dense materials that cut down the bulk

PREDICTION 1950 By 2000, supersonic planes cover a thousand miles an hour, but the consumption of fuel is such that high fares are charged. In one of these supersonic planes the Atlantic is crossed in three hours. Nobody has yet circumnavigated the moon in a rocket ship, but the idea is not laughed down.

Corporation presidents, bankers, ambassadors and rich people in a hurry use the 1000-mile-an-hour rocket planes and think nothing of paying a fare of $5000 between Chicago and Paris. Ordinary folks take the cheaper jet planes.

This was true—until the Concorde fleet ceased service in 2003. Research continues into the development of new supersonic passenger jets.

TRUE!

PREDICTION Hiller has made a study for a combination helicopter and airplane. This craft looks like an ordinary plane but has a large rotor overhead. The craft would use its rotor for taking off straight up, then as it began to use its main power plants to achieve forward speed, the rotor blades would fold and retract into the wing.

and weight. "The first plane," says one top-flight scientist in the atomic-energy program, "will be a military one—not commercial. Crews may be protected from gamma rays by simple shielding across the fuselage."

Suitable materials for building power units still are a headache. Nuclear piles aren't like high-temperature engines in which you just look for a harder or tougher steel. Fissioning atoms does strange things to most materials. Nitrogen comes out carbon; gold turns into mercury; sodium becomes magnesium; platinum turns into gold; glass comes out phosphorus.

"How would you like," Doctor Hafstad asked a group of airplane manufacturers recently, "to try to design aluminum plane engines which, during use, would slowly turn to cast iron or lead?" That's the kind of problem that keeps these scientists up at night.

PREDICTION 1953 The sound barrier itself—a mass of turbulent air hit by an airplane flying fast as sound waves—has now been conquered. What lies ahead is to transform today's uncertainty into tomorrow's reality.

What faster-than-sound flight means to the public is easy to guess. For example,

PREDICTION 1965

NEW DEAL FOR THE *HARRIED COMMUTER*

In air travel, efforts to design and place in service a supersonic airliner by 1970 are well known. Less familiar, however, is another development of almost equal significance, vertical takeoff and land planes (VTOLS). Some, using jet engines, blow the air down for takeoff and landing, and rearward for forward cruise. Others, using props, tilt the engines or tilt the wings. The most recent, the Bell X-22, uses four ducted props, tilting them up or down for forward or vertical flight. Success of any of them could lead to transports that could carry passengers from downtown areas, bypassing hard-to-get-to airports.

Roland Falk, chief test pilot of A.V. Roe and Co., Ltd., believes that within 50 years there will be a two-hour "commuter" service between London and New York. This will be done, he predicts, in a nuclear-powered aircraft, probably a delta shape, which will carry more than 100 passengers at speeds of around 2000 miles an hour and at heights 10 to 12 miles above the earth. The plane will make 10 trips a day and will land only when it needs servicing. Passengers, cargo and fuel will be carried to it by air taxis seating 25 people. Passengers will transfer between the two planes by means of a pressurized passageway.

The main plane, Falk contends, will not have windows because of the huge difference between atmospheric pressure inside and outside the cabin which would make window installation difficult. However, the passengers wouldn't have to be upset about this, because at a height of 10 miles they couldn't see much anyway.

THE CONQUEST OF SPACE

PREDICTION 1930 The scientific world is becoming rocket-conscious. The poles have been covered, the skies saddled, the mountains penetrated by engineers, nature's submarine secrets have been disclosed to the camera. Why not probe the interplanetary spaces?

Perhaps it is an exaggeration to talk of reaching the moon. Neither of the leading experimenters have any immediate hopes of doing so. But their preliminary tests have led them to make sanguine prophecies. Some day, the American asserts, man may send a rocket across those 221,614 miles to the moon. Before that, his German colleague believes, rockets may be speeding across the Atlantic bearing mail. Within four years, he asserts, there may be passenger rockets.

PREDICTION 1958 Though the United States Air Force refuses to deny or admit it, an elaborate space-platform program called Project Pied Piper (also known as Big Brother because of its watchdog implication) is reported well under way. A globe-circling reconnaissance satellite, orbiting at an altitude between 300 and 1000 miles, Pied Piper would carry TV cameras and transmitters, and radar or infrared scanning systems which could sweep all corners of the earth once every 24 hours, monitor global events such as weather, atom-bomb tests, undue massing of ship or aircraft fleets, and so on. Taken with telephoto lenses, pictures would carry detail as sharp and clear as an aerial photo taken at 5000 feet.

Target date for an unmanned version of Big Brother is estimated about 1960; manned version, about 1965.

PREDICTION 1958 An atomic rocket, already designed by the Atomic Energy Commission, carries a supply of liquid-hydrogen propellant that explodes into voluminous mass as it circulates through a uranium furnace in the nose.

This is merely a prelude to the day when man himself steps off into the void beyond the atmosphere—a day that, until last October 4, was considered a pure fantasy coddled by a few fanatics. Today, it is frighteningly close to reality. A leading Russian rocket expert has casually pre-

An artist's conception of the atomic-powered rocket. Liquid-hydrogen fuel would permit flights above Earth's atmosphere.

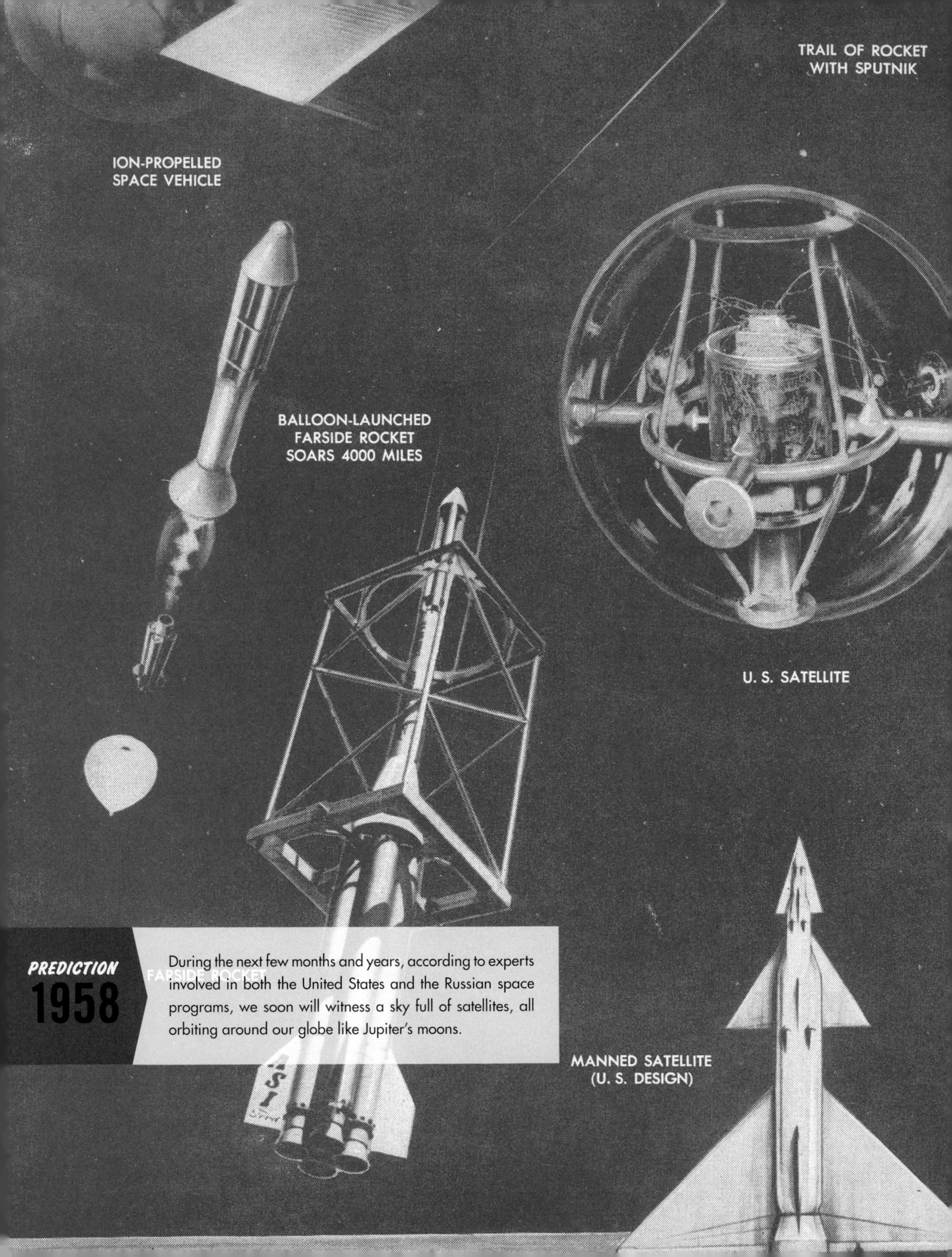

TRAIL OF ROCKET WITH SPUTNIK

ION-PROPELLED SPACE VEHICLE

BALLOON-LAUNCHED FARSIDE ROCKET SOARS 4000 MILES

U. S. SATELLITE

PREDICTION

1958

FARSIDE ROCKET

During the next few months and years, according to experts involved in both the United States and the Russian space programs, we soon will witness a sky full of satellites, all orbiting around our globe like Jupiter's moons.

MANNED SATELLITE (U. S. DESIGN)

dicted manned flight to the moon in three years; to Mars in 13 years.

One of the biggest problems of space travel is getting people back safely. Their vehicles, traveling at 18,000 to 20,000 miles per hour, would melt down to a glowing lump of tubing if they plunged into the atmosphere at such speeds. A winged-glider arrangement that would enable the vehicle to orbit through the atmosphere, slowing gently and finally landing like any aircraft, has been suggested. Actually, the first men to fly into space probably will get there not in satellites, but in craft similar to our present ramjet and rocket-driven high-altitude research planes.

PREDICTION 1958 This giant satellite-tracking camera is so accurate it can photograph a golf ball 60,000 feet in the air.

Most fascinating of all satellite prospects for the near future are the manned space stations, due within a dozen years according to experts. Such a massive satellite, capable of housing a crew of 50 to 75 men, might be a huge cylinder with a vast disk of solar batteries at one end or a doughnut-shaped affair in which men work and live in corridor-type rooms filled with instruments.

Advantage of launching spaceships from such an island in space is great. Fantastic speeds can be achieved since a rocket in space accelerates continuously at phenomenal rates as long as its fuel holds out. Using earth as a launching pad, most of the fuel is burned pushing the massive vehicle through the atmosphere.

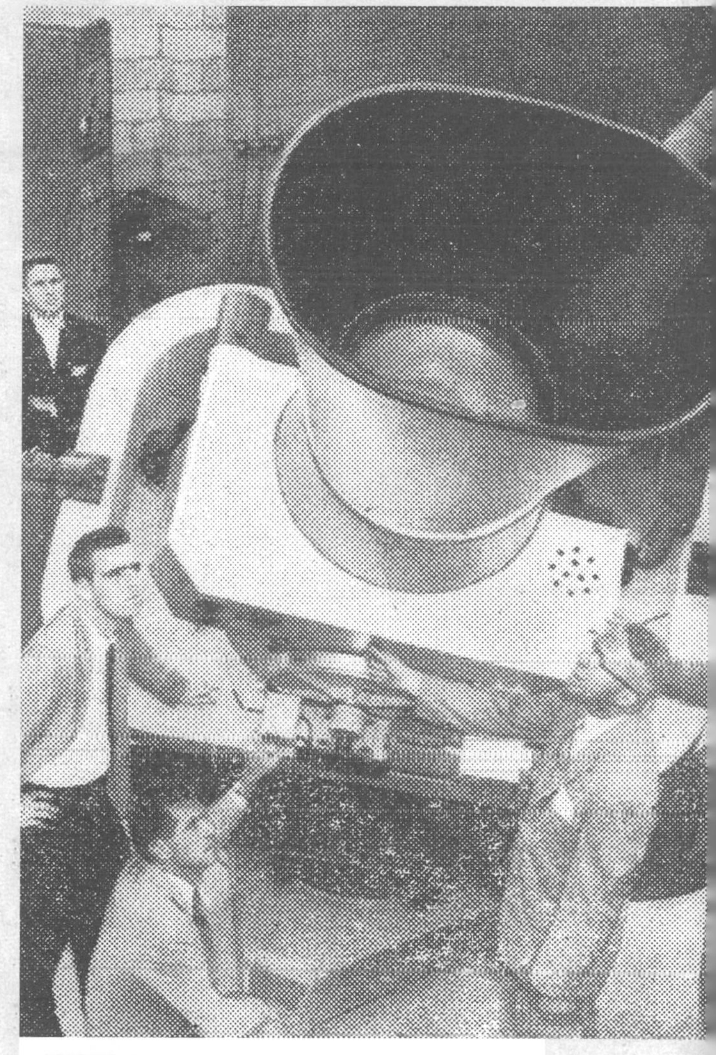

PREDICTION 1958 This giant satellite-tracking camera can photograph with remarkable accuracy.

Since friction is no problem in space, interplanetary spaceships need no streamlining. They will consist mostly of one huge spherical or cylindrical fuel tank. Small

AMERICANS WILL NOT BE THE FIRST *TO THE MOON*

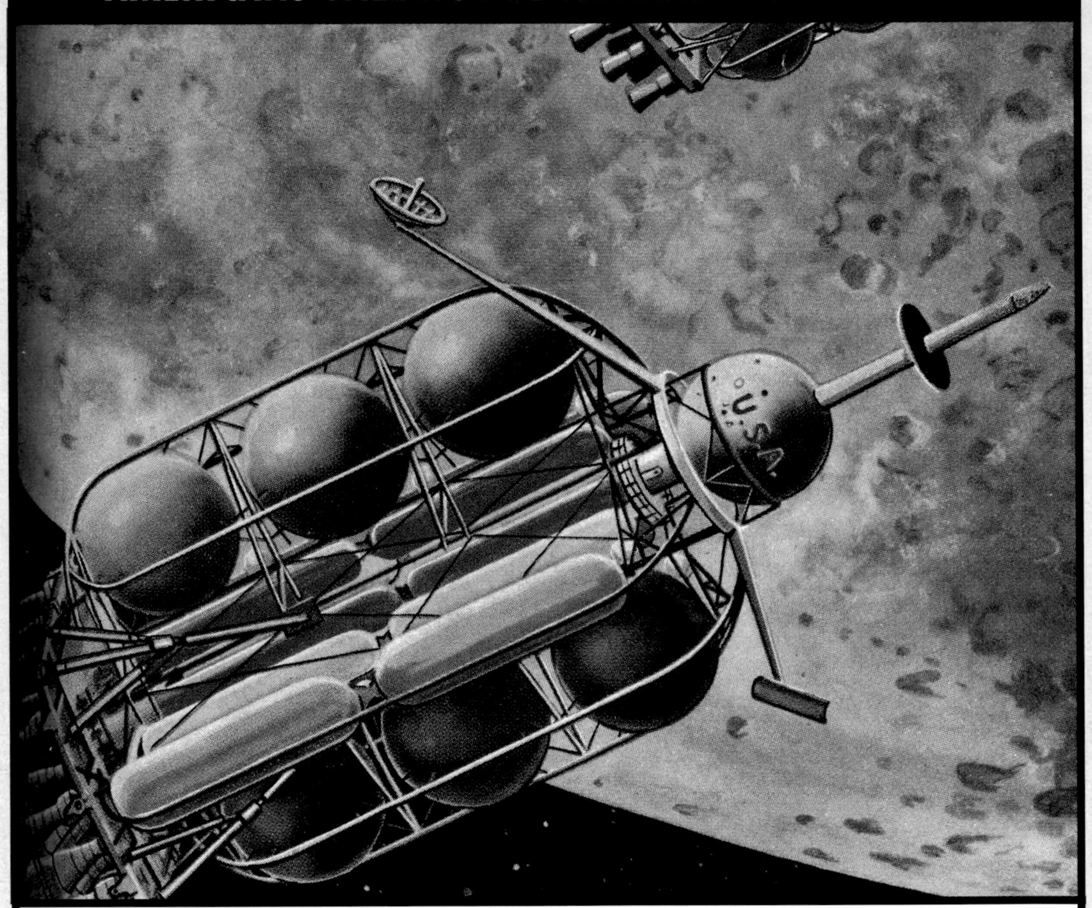

The first American trip to the moon will be launched not from earth but from a space station in orbit around our planet. It is possible even now, of course, to send a man to the moon the hard way; a single man cramped into a tiny shell with just enough fuel to reach the moon, land and fire himself back toward the earth. At the present time his chances of surviving would be slim, so the United States, at least, will ***not*** send a single man directly from earth. Our program is set, and does not include anything so wild and frantic. We will build an elegant space station to accommodate about 50 men, then set out in perhaps two personnel ships and a cargo ship. The Russians have a basic, relatively unpretentious moonship. The American ship is more elegant, larger and has a much bigger crew. This, perhaps, is the crux of the whole problem of the U.S. vis-a-vis Russia in space. The Russians think the stakes are nothing less than the cosmos. The Americans say, "Okay, the cosmos. But with safety, comfort, the dignity of man, showers in our space liners, big crew, togetherness, psychological adjustment, compatibility, friendship." The Russians whip something durable together and shout "Davai! Give, lads!" And off they go. I am certain in my own mind that the first spaceship will land on the moon within five years. And the way things are going at present, the men who emerge to put the first footprints into the ancient lunar dust will not be Americans.

By 1972, a radio telescope and astronomical observatory will be established on the moon. They may, under lunar conditions, discover things that would be impossible to find by observation from earth due to the gravity and atmosphere.

rocket motors and living quarters will be tied to the tank by metal girders. Spidery legs will provide something to land on. Taking off from the space station, they will move through the solar system like monstrous insects at six-figure speeds.

PREDICTION 1961 **Astronauts on long space journeys will change suits once a week**—and the suits will probably be made of paper. Rollin Gillespie, a scientist at the missiles and space division of Lockheed Aircraft Corp., says paper clothing seems the most practical solution to a number of problems. On a flight to Mars, for instance, it might take a year for a round trip, and this would involve a lot of laundering unless disposable clothing were worn. Paper clothing is also lightweight, warm, and absorbent of body fluids and odors, a major problem in cramped quarters on a long flight. A paper suit might look like long underwear or pajamas and feel like felt—porous, thick and soft. The Saturday-night bath might consist of no more than a "GI shower"—a sponge bath using a water-alcohol solution. New antibiotic deodorants would also help.

6
THIS UNFINISHED
WORLD

PREDICTION 1938

A working model of the interplanetary rocketport of the future at the moment the rocket departs.

Perhaps the greatest change men and women faced in the last century came from speed—the wonders of movement, of transportation. Cities led the way because they cannot work without the ready transport of products and services.

In 1900 the horse was the most common transportation device, but people had already envisioned car-dominated roadways and even moving roads. The twentieth century saw velocities change from the startling 60 miles per hour of 1900 to jet planes that topped 1,000 miles per hour. Urban movement and its compressions seemed the theme of the times.

Yet New York City's dwellers' persistently favorite place is not Times Square or the tourist attractions or museums. Instead, it's Central Park.

The most enduring tension in our vision of city life is between the closeness that kindles creativity, social life, and quick amusement, and the soft, quiet pleasures of the rural. Many studies reveal that when shown pictures of places to live, people prefer homes midway between bodies of water and either mountains or forests or both. This echoes where we evolved. Our ability to move from land to water, forest to grassland, mirrors our survival strategies then: avoiding predators and changes in weather, and looking for food.

The sense of wonder that marked the twentieth century should animate us in the twenty-first. Many ideas from that time can work even better now. Thomas Edison proposed concrete houses for fire safety and durability, but they never caught on because people found them ugly. But if well done, they can obviate damage and enhance neighborhoods.

Even things that got ridiculed back then happened anyway. Famously, in *The Graduate*, the advice, "Buddy, I've got one word for you: *plastics*," elicited derisive laughter, but in fact plastics now appear everywhere in our lives. Weather control still eludes us, though often cloud seeding does seem to work. The agenda set by the age of techno-wonder is still unfinished.

What does this portend for the next century? Clearly we have come a long way from unblinking wonder at technology. We distance ourselves from the top-down social engineering doctrines that often drove the optimism of a century ago. Imagining how science and its handmaiden, technology, could affect society now more often employs the self-organizing principles popular in biology, economics, artificial intelligence and even physics. The twentieth century (or the slangy "TwenCen") has been the century of physics, just as the nineteenth was that of mechanics and chemistry. Biology will rule the twenty-first.

Still, grand physical measures still beckon. We could build a sea-level canal across Central America, explore Mars in person, use asteroidal resources to uplift

PREDICTION 1939 The force of gravity on Mars is one-third of that on the earth and therefore we would seem to be three times stronger. Ten-foot high jumps, forty-foot long jumps and lifting 300-pound weights would all be possible. This would be of great advantage if any creatures of Mars attacked us, for we should be the equal of three of them. And if there were too many of them to fight, we should also be able to run faster than they can.

the bulk of humanity. Siberia could be a fresh frontier. All will respond to new technologies. All are possible.

In 1900 world population was 1.6 billion. Now it's 6.8, and most estimate we're headed for 10 billion by 2050. That enormous challenge demands new approaches, fueled by visions of futures that are utopian, even if we don't get quite all the way there. The magazines of the twentieth century show us how to dream, with constraints.

Our great problems all involve new technologies. But no one can achieve anything that he or she does not first imagine.

PREDICTION
1925

With the upper air conquered and the globe girded by flyers; with both poles of the earth visited and the deserts spanned by auto busses; man, looking for new worlds to conquer, has turned again to the bottom of the sea.

STORMS ARE UNDER CONTROL

PREDICTION 1905 The farmer may well conduct the whole of his operations by machinery and the application of scientific principles before the millennium. Old Sol still sheds his luminous rays which are utilized to their full efficiency by glass enclosures, but he is assisted by powerful electric lights and vast systems of radiators. Unless the California rainmaker has a better system, storms will be produced by firing mortars and shells and dissipated by the same means, on occasion. The whole system will be under the control of one or two operators and the farmer can tilt back in his chair, elevate his feet, smoke, and watch the grain grow. We predict that an American farmer will be the first to have a plant of this kind.

PREDICTION 1905 Electric farm of the future may come about as early as 1930 A.D.

PREDICTION 1911 In his recent presidential address to the Institution of Electrical Engineers, in London, England, S.Z. de Ferranti, a man whose practical accomplishments have helped to make electrical history, unfolded what he himself termed "visions of the future," a story of the coming electrical age, when all power for the production of heat, light and mechanical energy will be generated electrically in a few great central stations distributed throughout the land. Electric smelting would supplant present-day methods, electro-chemical industries would prosper exceedingly, and even the rainfall would be controlled electrically. "Of course, there are many things which at present stand in the way of realizing such a scheme as I have outlined," declared Mr. de Ferranti, "but the more I have considered these ideas in detail, the more certain am I of the fundamental soundness underlying them, and it is only a matter of time before such a scheme is carried out in its entirety."

PREDICTION 1911

FOG DISPELLERS WILL PREVENT COLLISIONS

New York Harbor might look like this during a fog, if electrical fog dispellers were used to clear a channel for ocean-going steamships. The time will come when such devices will be put into operation in New York and along the Chicago River and the Thames. Possibly in the far-distant future, all of our large cities may have their Bureaus of Mists and Fogs, enabling us to disperse at will the occasional fog and the perennial smoke which envelops us.

PREDICTION 1950 One of the more remarkable electronic machines of the Year 2000 is one that will predict the weather with an accuracy unattainable before 1980. It is a combination of calculating machine and forecaster. The calculator solves thousands of separate equations in a minute; the automatic forecaster carries out the computer's instructions and predicts the weather from hour to hour. In 1950, meteorologists had no time to deal with the 50-odd variables that should have been mathematically handled to predict the weather 24 hours in advance.

With the use of this machine, it is easy to spot a budding hurricane off the coast of Africa. Before it has a chance to gather strength and speed as it travels westward toward Florida, oil is spread over the sea and ignited. There is an updraft. Air from the surrounding region, which includes the developing hurricane, rushes in to fill the void. The rising air condenses so that some of the water in the whirling mass falls as rain.

PREDICTION
1963

Weathermen believe they can knock out a hurricane by cooling down the primary source of its awesome power—a heat chimney, or tower (in red), located in the storm's right front quadrant.

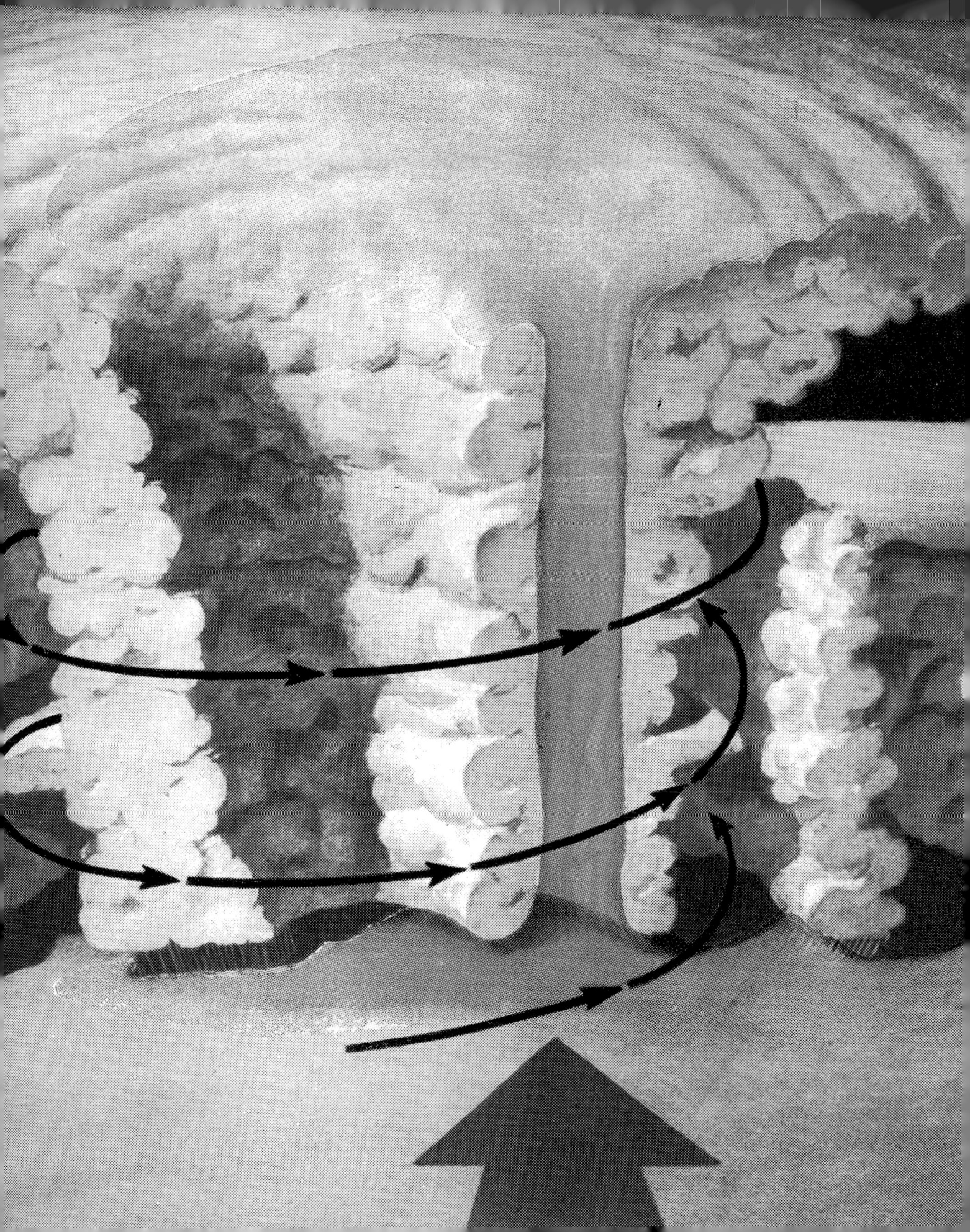

FLOATING PLASTIC DOME COULD COVER STADIUM

PREDICTION 1942 Research may allow us to do incredible things with super-lightweight materials. The starting point may be the "aerogel" compounds now used mainly as gas or moisture absorbers. These substances are composed of innumerable minute bubbles like foam, stuck together to form a solid mass, some of them are 20 times lighter than water.

Imagine a house built of such bubbles instead of heavy brick, wood, glass, and plaster. Its cost would be little because there would be little material needed in the first place, and no heavy materials to transport and erect. Instead of expensive foundations, simple wind anchors would hold the house in place. The walls could be formed in place with a spray gun, and the roof arched over, integral with the walls.

Our bubble house is automatically sound- and vermin-proof, insulated against

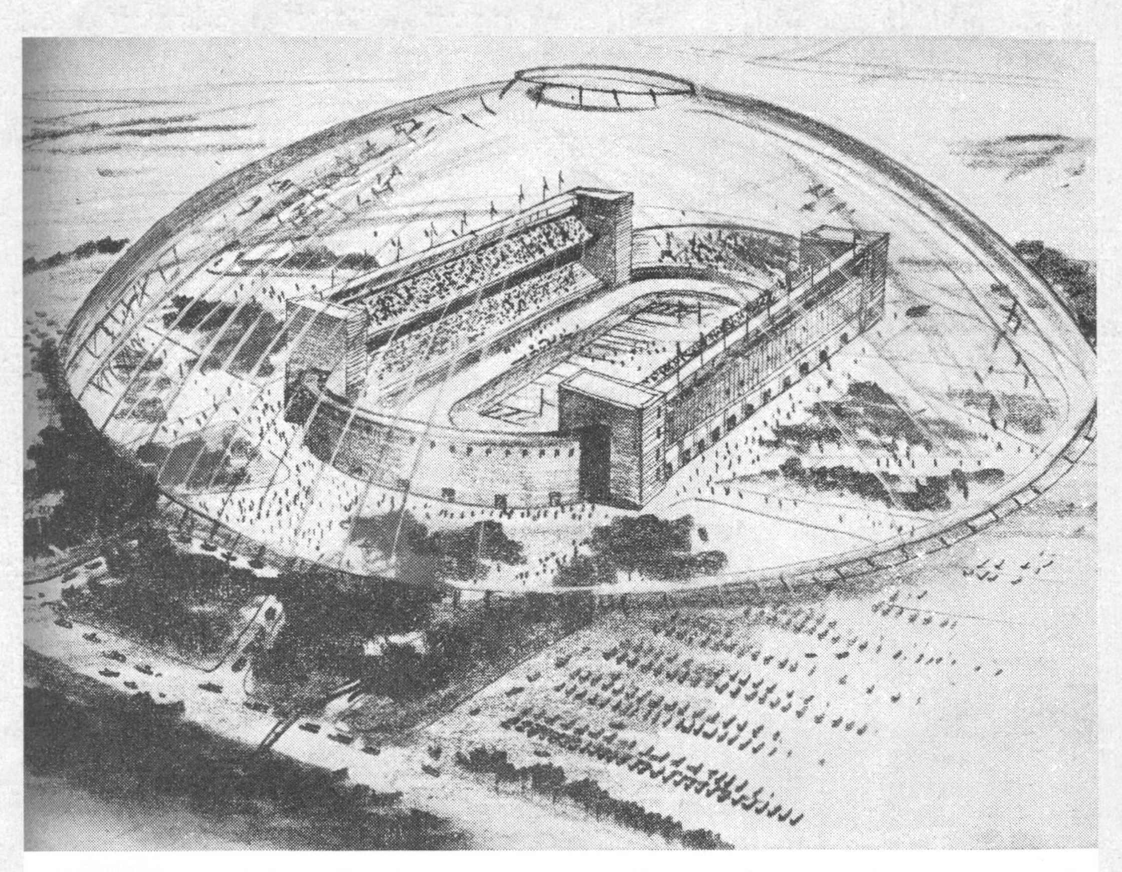

Dome of plastic pillows has air vent at top covered by a second canopy. Cool air enters at top and flows down to force warm air out the edges which would be 10 or 12 feet from the ground. A dome of this type could cover a big stadium or ball park for $200,000.

extremes of temperature, and has no heavy elements to collapse in an earthquake. It is fireproof. Part or all of the walls could be left transparent so that with draperies and shades you could arrange your windows to suit your convenience, and change them when you wish. With the chief use for cheap lumber eliminated, our forests would become parks and game preserves instead of useless cut-over land.

PREDICTION 1953 **Plastic windows filled with helium and joined to make a floating umbrellalike dome may cover stadiums and baseball parks of the future.** Prof. Ambrose M. Richardson, University of Illinois architect, believes it may be eventually possible to cover entire cities, with industrial buildings and factories located outside of the dome. The dome would be anchored by cables with no interior support except

Twelve men could lift an aerogel-constructed house that would weigh ten tons if made of materials in use today.

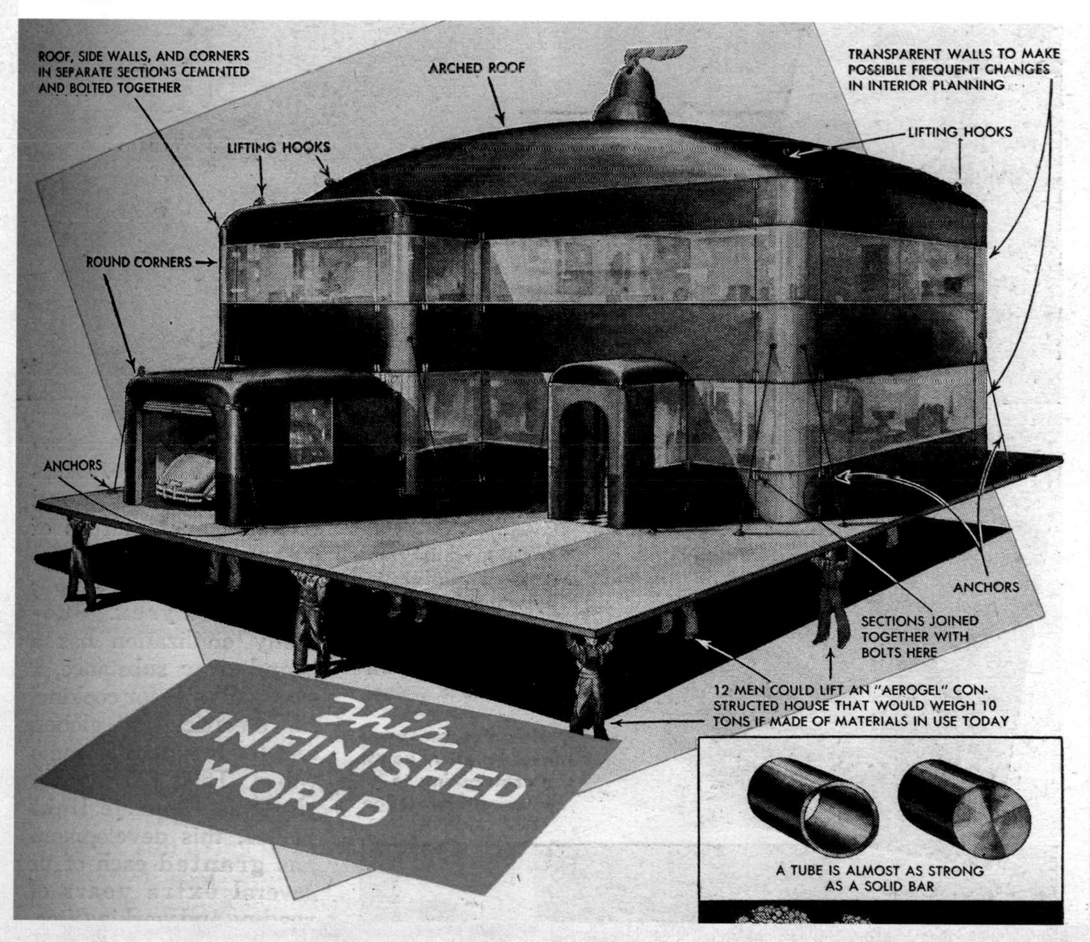

buoyant helium to hold the pillows aloft with edges of the dome 10 feet off the ground. Richardson envisions homes under the dome built without roofs and with lightweight walls for privacy. Even on overcast days, he predicts, solar radiation would provide adequate heat. Supplementary radiant heat could be arranged for cool evenings.

COLORS TO LIFT ONE'S SPIRITS

PREDICTION 1914 **Daylight glasses, or spectacles which when used under artificial light** will permit the entrance to the eye of only a daylight spectrum, excluding all other rays, are a possibility of the near future. The uses of daylight glass would be manifold, particularly in color printing, artificial-teeth manufacture, the selection of diamonds, matching of fabrics, and similar things where the perception of colors as they appear in daylight is absolutely essential. Glasses of this nature would thus allow many institutions to operate on dark or cloudy days without interruption, a thing now impossible. Such an invention would allow a person, visiting an art gallery at night, to obtain the same effects as if he were viewing the paintings under the most favorable circumstances.

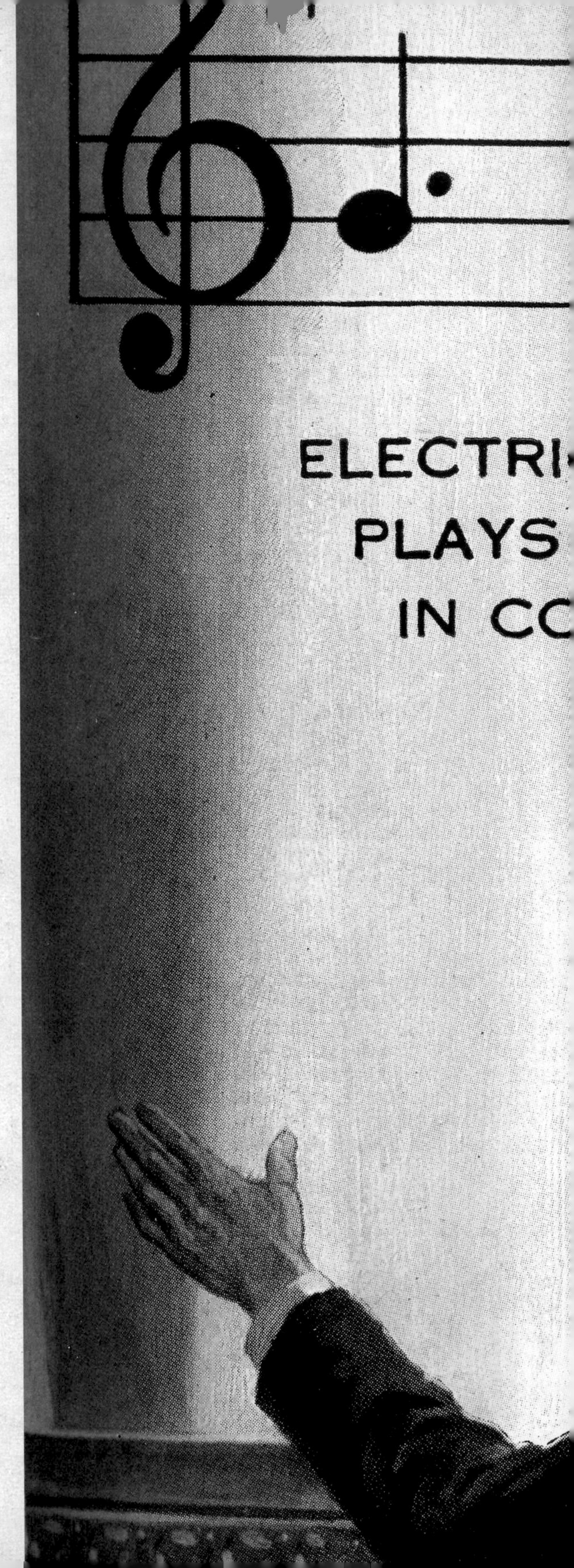

Degrees of emotion from placid contentment to intense manifestations are vividly expressed by color parts used in connection with orchestral music in some of the greatest concert halls of the world. This image depicts a color phase as it appears to the spectator-listener.

PREDICTION
1915

PREDICTION 1924 The audience sat in hushed and wondering expectancy within the darkened theater. Without accompaniment of sound, soft color suddenly glowed upon the screen. Slowly it moved into definite form, its modulation of figures evolving in majestic sweeps. Its hue deepened and then melted radiantly into iridescent crimson, and from the restless, ever-changing shapes a slow rhythm was born. It grew and blossomed, a symphony of light, plastic and mobile.

The "clavilux," as Thomas Wilfred, the inventor, has named the organ, opens the door to a new art, the expression of moving color and form, which the artist-craftsman believes is destined to take a place as a sister of music and sculpture. It has long been the vision of dreamers; Mr. Wilfred has actualized the dream and provided the instrument that visualizes it. "Light music is as much an art as painting, sculpture or dancing, and is a wonderful medium for artistic expression."

PREDICTION 1938

Artist's sketch of a nightly spectacle in which water, flame, color, and sound will be combined in almost incredible proportions. At the touch of a control, fountain jets and towers of flame 150 feet high will spring from hundreds of water nozzles and gas burners to form awe-inspiring designs.

"I confidently predict that a few years will place 'light concerts' beside symphony concerts, the opera and the movies. And every home will have a color organ, as every home now has a piano or a phonograph," says Wilfred.

The creation of a new art! A thousand or more color forms, constantly changing in tone and in hue, are projected on screen at the touch of an artist on an organ.

PREDICTION
1924

PREDICTION 1939 In the not-distant future, the home may well be equipped with "mood control," which is made possible by newly developed light sources. It's possible that people will suit the light and color of their rooms to their moods. These new-type lamps produce colors of warm white, daylight white, gold, red, blue, pink and green. It's up to the psychologists to figure out the proper combinations of colors to lift one's spirits, when they are down, with a flood of brilliant light, or subdue a sense of excitement with soothing mellow light.

PREDICTION A "people sniffer" used in Vietnam spots the enemy by detecting traces of ammonia in body sweat, and is a forerunner of more sophisticated artificial noses.

PREDICTION 1957 By A.D. 2000, ways will be found to transmit information to the brain in such a way that loss of sight and hearing will not restrict one's activities in any way. And the senses of people with normally good vision and hearing will be strengthened; for instance, it will be possible to see in total darkness.

CLOSELY BUNCHED PEAKS REPRESENT MOST VOLATILE VAPORS DRIVEN OFF FAST AT START OF ANALYSIS

TYPICAL TALL PEAK INDICATES HIGH CONCENTRATION OF ONE PARTICULAR ODOR CHEMICAL

THEY'LL KNOW YOU BY YOUR SMELL PRINT

PREDICTION 1935 We already hear and eventually will see by radio, says Dr. Alfred N. Goldsmith, consulting engineer of the RCA Manufacturing company. He suggests that in the remote future radio may appeal to man's other sense, such as taste, smell and touch. "Telegustatory Broadcasting," as he calls it, may make it possible to taste a fine brand of coffee by radio. The transmission of smells perhaps would be easier, he thinks, since the "telolfactory" receiver would need only to spray into the air a duplicate of the odor transmitted. Too, there is the very remote possibility of transmitting three-dimensional replicas of objects which might be touched.

PREDICTION 1968 Olfactronics, the new science of smells, is already making a name for itself. Its best-known achievement to date is the "people-sniffer," developed by General Electric, that's being used successfully to detect enemy troops hidden in the jungles of Vietnam. An olfactronic bomb detector has been developed to smell out explosives hidden with murderous intent aboard airliners. In the future, precise smell analysis will help doctors to diagnoses diseases. Sniffers will be used in industrial process control and even by electronics servicemen to identify malfunctioning components.

PREDICTION 1968 There is a bright future for olfactronics in medicine. Since human odors have their origins in biological processes, changes in odor signatures can be used to detect biological malfunctions. Physicians already use their sense of smell in diagnosing diseases, and about 40 different medical conditions are known to have associated odors. But the usefulness of this tool depends on the personal experience of the doctor.

SCIENCE

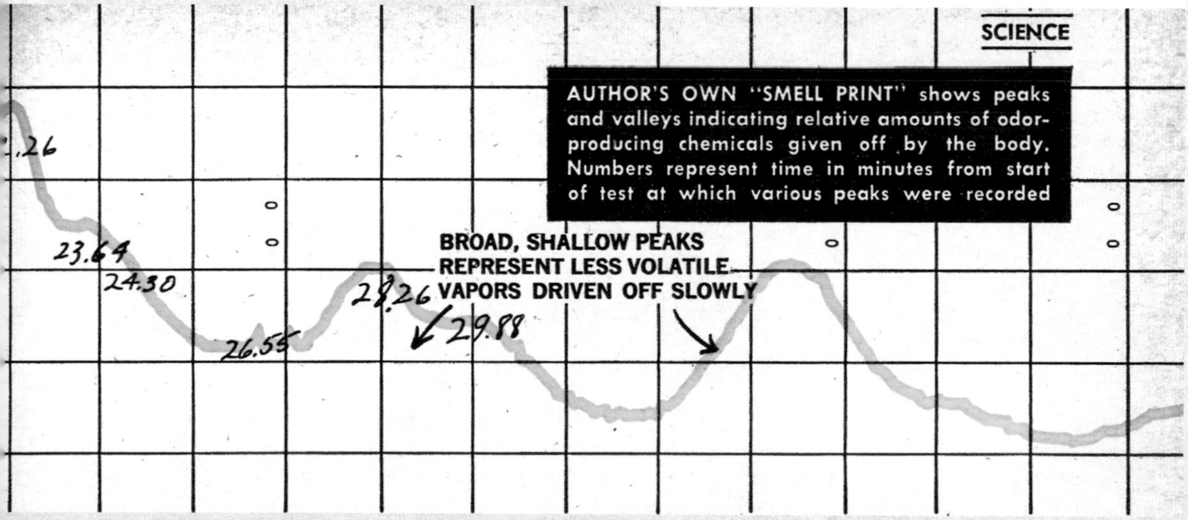

AUTHOR'S OWN "SMELL PRINT" shows peaks and valleys indicating relative amounts of odor-producing chemicals given off by the body. Numbers represent time in minutes from start of test at which various peaks were recorded

BROAD, SHALLOW PEAKS REPRESENT LESS VOLATILE VAPORS DRIVEN OFF SLOWLY

.26
23.64
24.30
26.55
27.26
29.88

Until only a few years ago, smells were beyond scientific analysis. Odors consist of such small quantities of vapor in the air that instruments were not sensitive enough to check them out. But this has all changed. Apparatus such as that used at the IITRI olfactronics laboratory is now capable of detecting many substances 1/100th to 1/10,000th as concentrated as those noticeable to the nose.

Eventually, it is expected that olfactronic instruments, more dependable and expert than a doctor's nose, will be among the most sensitive of medicine's diagnostic tools. They may be able to spot some diseases even before people know they are sick. And they will be used for disease prevention—in schools, say, where the unsuspected carriers of airborne diseases will be detected by monitoring the air.

PREDICTION 1939 This "mechanical nose" can detect carbon monoxide in a manhole. Future instruments may be even more sensitive to smell.

"DEATH RAY" CARRIED BY SHAFTS OF LIGHT

PREDICTION 1924 Acting as a carrier for energy made up of high-tension electric currents, a new ray invented by H. Grindell-Matthews of London, claims to be able to destroy aircraft. The beams render the air highly conductive to electricity.

The ray has been tested to stop the operation of automobiles, and additionally a quantity of gunpowder exploded when the beam was directed on it from a distance of thirty-six feet. Where attendants have crossed the path of the ray during testing, they were rendered unconscious by violent shocks. This discovery may result in the development of entirely new fields and methods of warfare.

Should the power of the ray be harnessed, future wars may find air fleets combated by barrages of deadly light beams.

PREDICTION 1959 All over the world, scientists are making big strides in their efforts to harness the H-bomb for peaceful ends.

Scientists estimate that sometime within the next 10 to 20 years the first full-scale, power-producing fusion reactor will go into operation. Even this first crude reactor probably will have a power output comparable to the huge hydroelectric plant at Hoover Dam.

That moment, if it comes, will be a pivot point of history. Nations need never fear

Since the airplane has become a factor in commerce, the question of suitable landings within city areas has grown in importance. One plan calls for an immense stage to be erected on top of four skyscraper towers, to span 1,400 square feet. The entire platform can handle 80,000 passengers and 30,000 tons of freight yearly.

PREDICTION
1926

PREDICTION 1923

Current drawn from the air—neither lightning nor the charge of the earth, but said to have a sort of cousinly relationship with them—could supply all our energy needs by unleashing and harnessing the power of the atom. Here an experiment generates the most terrific heat known, estimated at 50,000 degrees.

that their power sources will run dry. For a time, obviously petroleum will remain as a source of mobile power, and coal will continue to provide industry with heat. You can't change a way of life overnight. But eventually, according to the experts, these will give way to stored electrical power derived from the fusion reaction.

At that supreme moment, man can look ahead through the halls of time for literally billions of years and still see a plentiful supply of fuel. Billions of years! When new kilowatts are needed, they simply will be plucked from sea water. Inexhaustible and inexpensive power on earth will give man the means of navigating the stars.

PREDICTION 1966 Billions of watts of electricity may be supplied to cities of the future in the form of high-frequency radio waves transmitted through underground pipelines of foam plastic. Stanford University engineers say the method would have these advantages over conventional power lines: It could carry bigger loads, would be shielded from attack by the weather or an enemy nation and would eliminate unsightly towers.

PREDICTION
1959

FOUR APPROACHES IN THE ATTEMPT TO *CONTROL FUSION*

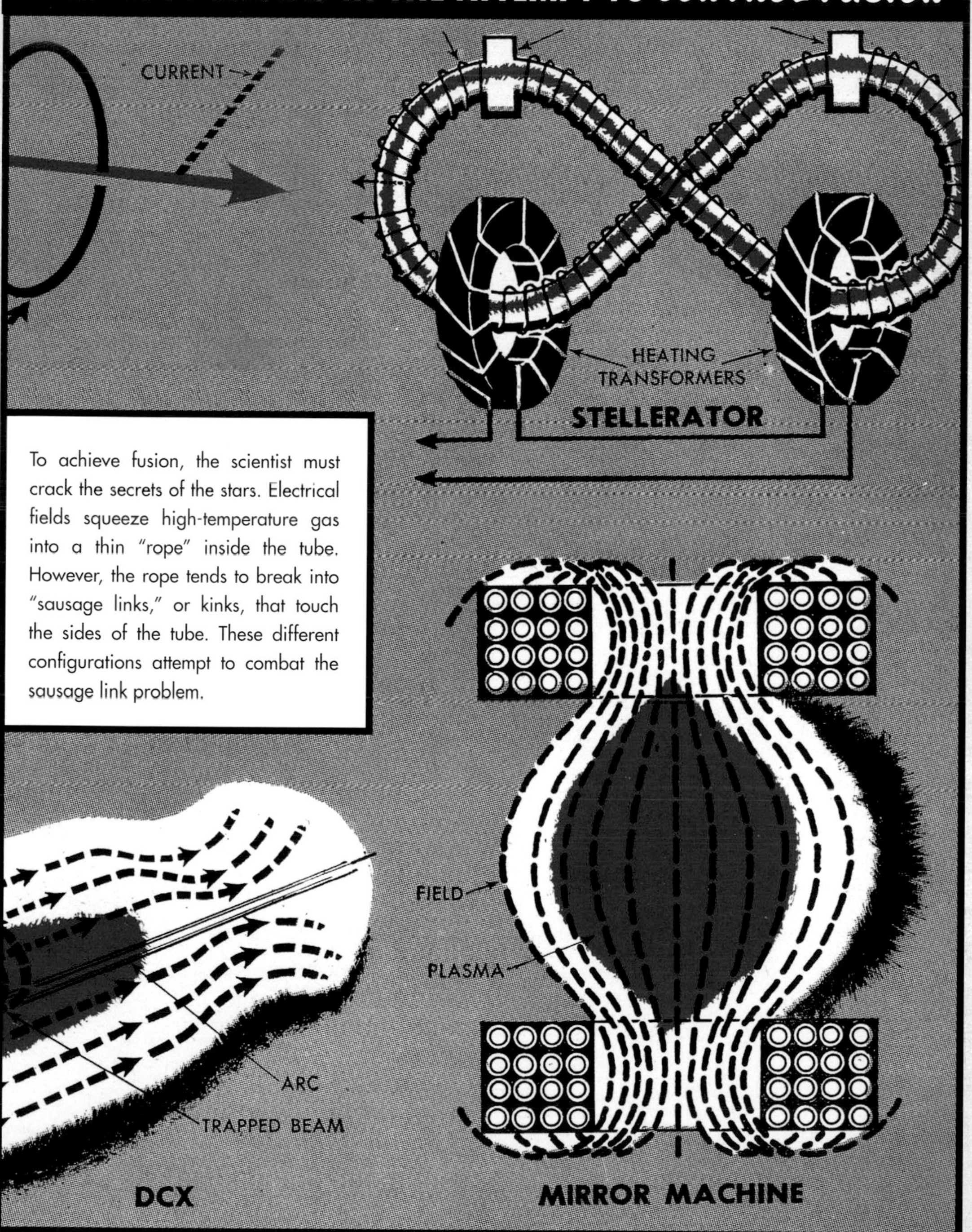

To achieve fusion, the scientist must crack the secrets of the stars. Electrical fields squeeze high-temperature gas into a thin "rope" inside the tube. However, the rope tends to break into "sausage links," or kinks, that touch the sides of the tube. These different configurations attempt to combat the sausage link problem.

AROUND THE WORLD IN HALF AN HOUR

PREDICTION **1929** Prof. Albert Einstein's new equations assume gravitation and electricity are the same thing. While the theory itself is almost incomprehensible to the layman, it opens the field for fascinating speculation. Since it is possible now to insulate against electricity, it follows that, if electromagneticism and gravitation are the same thing, it may be possible some day to insulate against gravity. If that ever comes to pass, motorless aircraft may ride the skies, people may step out of skyscraper windows without falling to the ground, and a trip to the moon becomes theoretically possible.

PREDICTION **1938** Looking into the future of invention, a British engineer predicts that man may some day discover how to control gravitation—a revolutionary inventive step that would have a greater effect on human living conditions than any previous discovery. He foresees steam power for huge aircraft, and the transmission of power through the air. Airplanes could be flown without fuel, drawing their power from beams of energy radiated along their routes.

PREDICTION **1945** An engineer's idea of an experimental rocket plane, comparing closely with present-day airplane shapes modified for high speed. All steering is accomplished by the directable jet nozzle.

Suggestions for bullet-shaped rocket planes, without wings, are impractical for carrying humans. The terrific acceleration required to shoot a bullet-shaped rocket up out of the atmosphere would kill anyone on board. No one could ride a V-2 rocket and live. Such a rocket, too, could not slow down enough for a safe and comfortable landing. Thus, future space ships will have wings, control surfaces, and conventional landing gear. Acceleration will be held to a maximum of three Gs, three times the pull of gravity, for comfort.

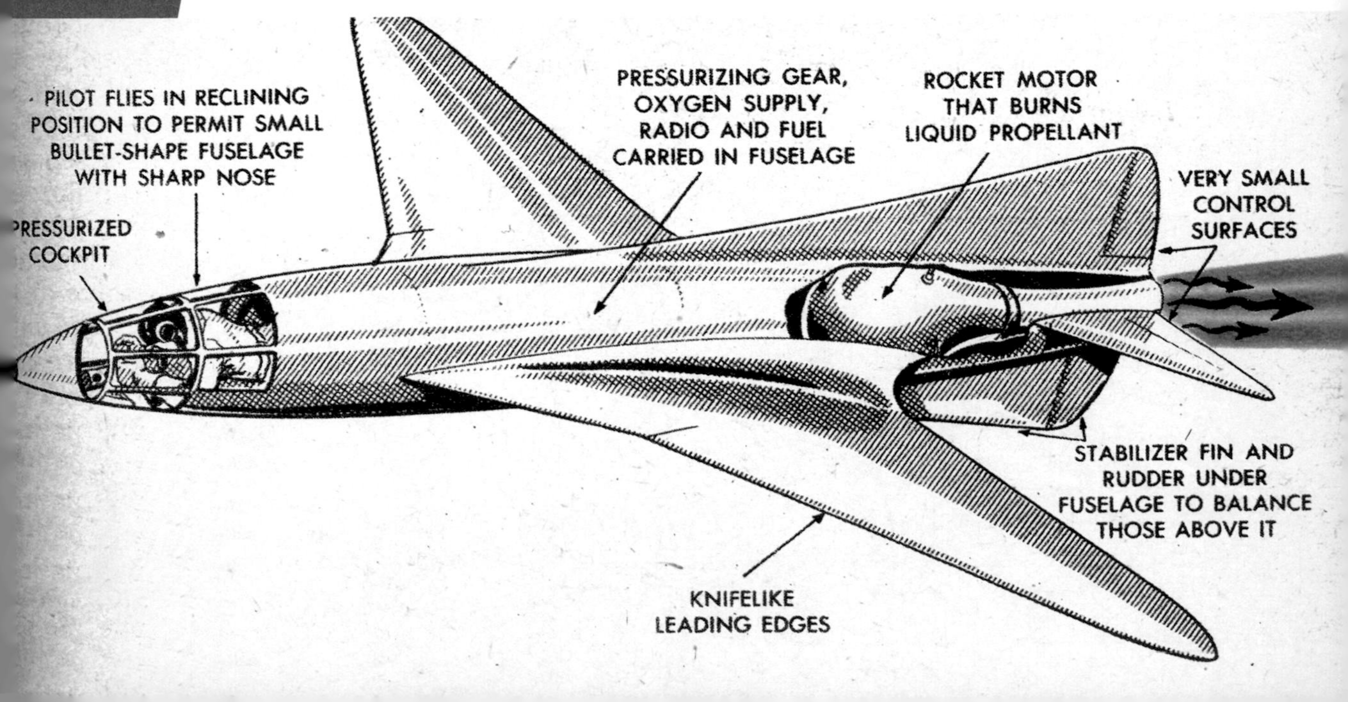

The airship of the future! A motorless craft flying through space is insulated against gravity.

PREDICTION
1929

PREDICTION

Twin rocket nozzles, spewing an invisible hurricane of steam, power this personal rocket-pack.

PREDICTION 1945 Aviation designers are looking forward to creating within the "visible future" a rocket ship that would fly at the rate of 100,000 miles per hour.

Hall Hibbard, vice-president and chief engineer of Lockheed Aircraft Corporation, states that the basic principles of such a spaceship are understood and the problems that remain are minor in comparison to the ones already solved.

It is conceivable that such an airplane, moving outside the atmosphere, could fly all the way around the world in half an hour. Provided that it could carry enough fuel and a livable atmosphere for its occupants, it could travel to the moon, around it and return to earth between daybreak and dark.

PREDICTION 1964 In 10 years, maybe less, some of you will be up here flying—as a Rocket Man.

This man rocket was developed by Bell Aerosystems, a division of the Textron Corp., and they predict that m an-rockets—and their high-rocketing wearers—will be as common as helicopters in the decade to come, and even more versatile.

Man-rockets will provide firefighters with seven league boots—whisk them, in seconds, to within reach of a skyscraper's loftiest holocaust. Outfitted with back-packed rockets, surveyors and construction men will find no hill too high to climb, and no river too wide to cross. No longer will military patrols need to pick and probe through mine fields—they'll simply rocket over them.

For sportsmen, rocket power will open inaccessible back country, fly anglers to the hottest fishing spots, and nimrods into the world's roadless big-game haunts. They may even fly commuters over the bumper-to-bumper traffic.

A new, more efficient fuel promises to give Bell's man-hefting rockets 50 times their present range, up to perhaps 10 miles of free flight at 60 mph. or better. The new fuel catapults man-rockets from the realm of limited-range prototype to workaday reality.

Historically, it's also a milestone. In little more than three years since man's first back-packed free flight, man-rockets may fulfill man's dream to fly like a bird.

COLONIZING THE OUTER PLANETS

PREDICTION 1927 How would you like to live on an earth that had eight moons wheeling overhead? Some day one of the neighbors in our universe—Jupiter—may support a race of people who will experience that unusual novelty. The possession of eight moons will not be the only distinction its people can boast, for their years will be as long as nearly twelve of ours—4,332.58 days to be exact.

If there is any place in our universe for human life to exist, outside the earth Venus or Mars come nearest to it.

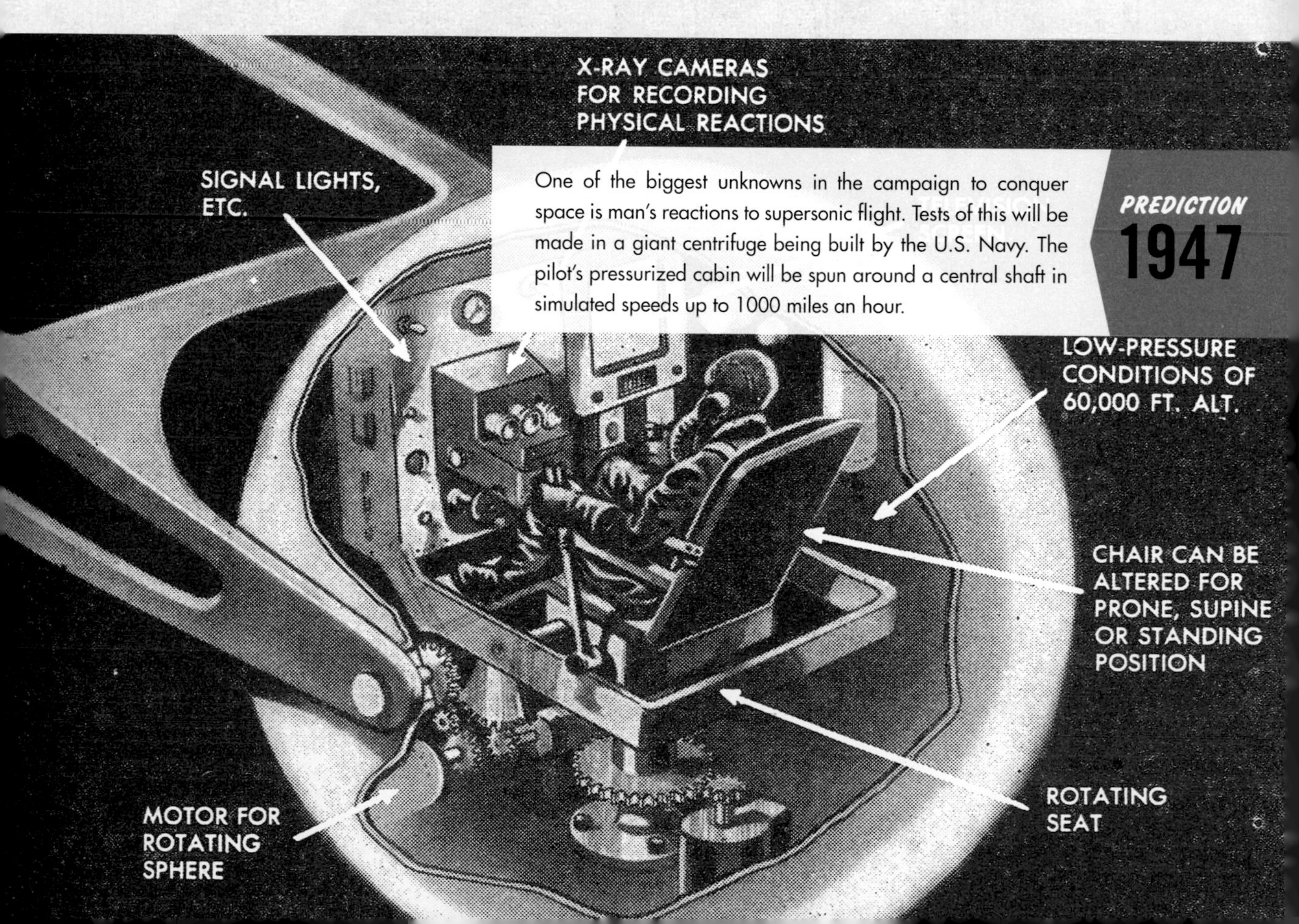

One of the biggest unknowns in the campaign to conquer space is man's reactions to supersonic flight. Tests of this will be made in a giant centrifuge being built by the U.S. Navy. The pilot's pressurized cabin will be spun around a central shaft in simulated speeds up to 1000 miles an hour.

PREDICTION 1928 Construction of man-carrying rockets capable of crossing the millions of miles of outer space and reaching the planets is scientifically possible and may eventually come, according to Fritz von Opel, the German automobile builder who, within a few months, has produced a rocket automobile and a rocket airplane.

One problem that remains to be solved is whether the human body could withstand the terrific strains involved in being shot into space at such speeds. In experiments with a rocket car mounted on railway wheels and operated on a track, Opel placed a cat in the seat to see how it would withstand the strain. The car, however, blew up and the cat was killed, so nothing was learned.

PREDICTION 1947 Extremely little fuel would be needed for travel to the planet Venus, considering the distances involved. Instead, the rocket captain would take advantage of the gravitational attractions of the sun and planets and would be drawn steadily toward his goal. Venus travels faster than the earth and so the space ship would take off while Venus was still far behind. Venus would overtake the earth and pass the space ship, then the space ship would catch up with it.

The crew of the ship would have no inkling of what they would find until after they had plunged through Venus' cloud

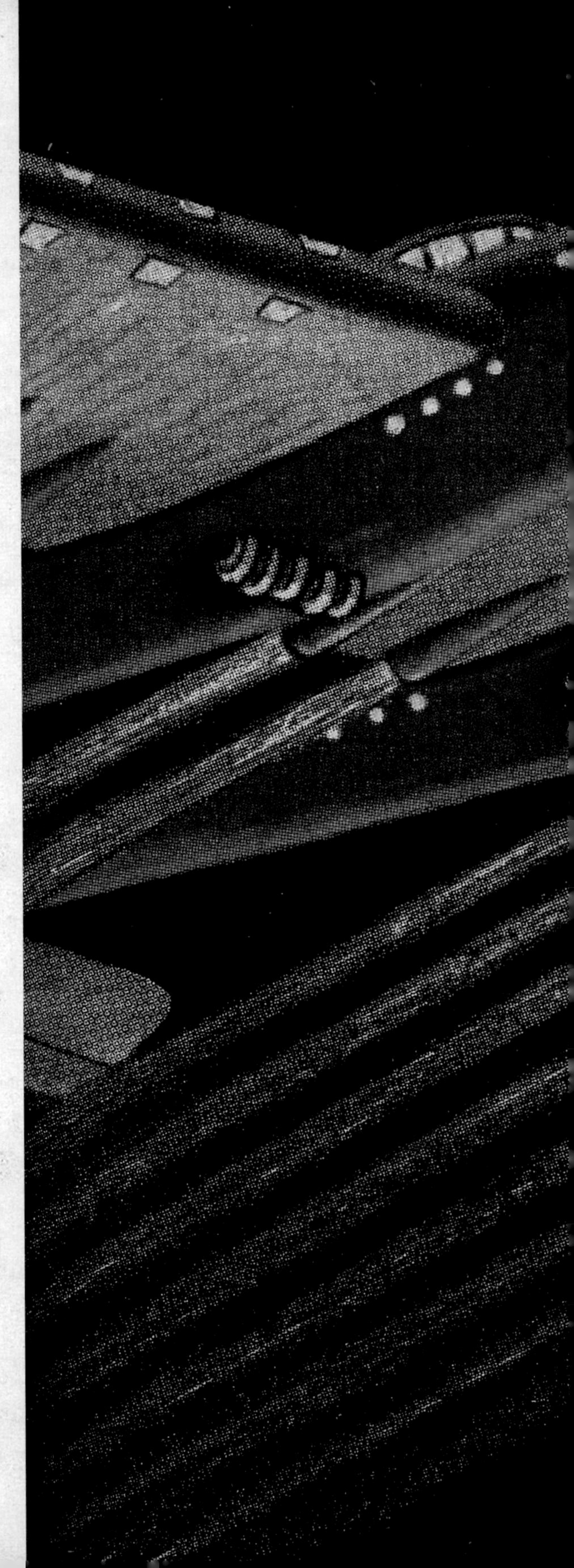

A rocket plane could fly out in space beyond the atmosphere, where it would have for neighbors the northern lights and occasional meteors.

PREDICTION 1928

curtain. If they landed safely they would have to begin manufacturing, at once and from whatever materials they could find, an atmosphere similar to ours. They would have to refine sufficient fuel for their return from the elements available. They would need to work out the navigation problem of returning home against the attraction of the sun. They would need derricks and other apparatus for re-aligning the craft for launching. They might even need a ground handling crew.

Obstacles like these make interplanetary travel appear almost as fantastic today as the idea of talking great distances over metal wire would have seemed several centuries ago. But the miracle of the telephone is old stuff now and in 1947 scientists only say that they don't yet command the magic to make space travel practical.

PREDICTION 1951 In years to come, a giant "doughnut" 200 feet in diameter may be traveling constantly around the earth 1075 miles up in the sky. That's the prediction of Dr. Wernher von Braun, credited with inventing the German V-2 rocket and now an Army rocket expert. The space station would be carried to its orbit in a three-stage rocket, and assembled by men in pressurized suits. They would then live in the outer rim of the doughnut, kept in place by a synthetic "gravity" produced by its rotation around the hub.

PREDICTION 1952 The time is coming when mankind will have to reconstruct the solar system. Fritz Zwicky, professor of astrophysics at the California Institute of Technology, says that we may have to rearrange the planets and in some cases rebuild them to fulfill our future requirements for living in space.

Not even Mars could be colonized on a large scale at present because it lacks a suitable atmosphere. It may pay to send Mars off on a tangent past some other planet to draw off atmosphere and then return the red planet to its present orbit.

To make one of the large outer planets habitable would be an even more tremendous task. The planet would have to be broken apart or shrunk to about the size of earth to provide a comfortable surface gravity. It would have to be endowed with a useful atmosphere. Finally, it would have to be moved closer to the sun, at about earth's orbit, where it could absorb enough solar radiation for our needs.

These are fantastic ideas, but are they absurd? Not necessarily, Professor Zwicky believes. He points out that unlimited power will be available when nuclear fusion is achieved and that no fundamental principle seems to stand in the way of using nuclear fusion on a large scale.

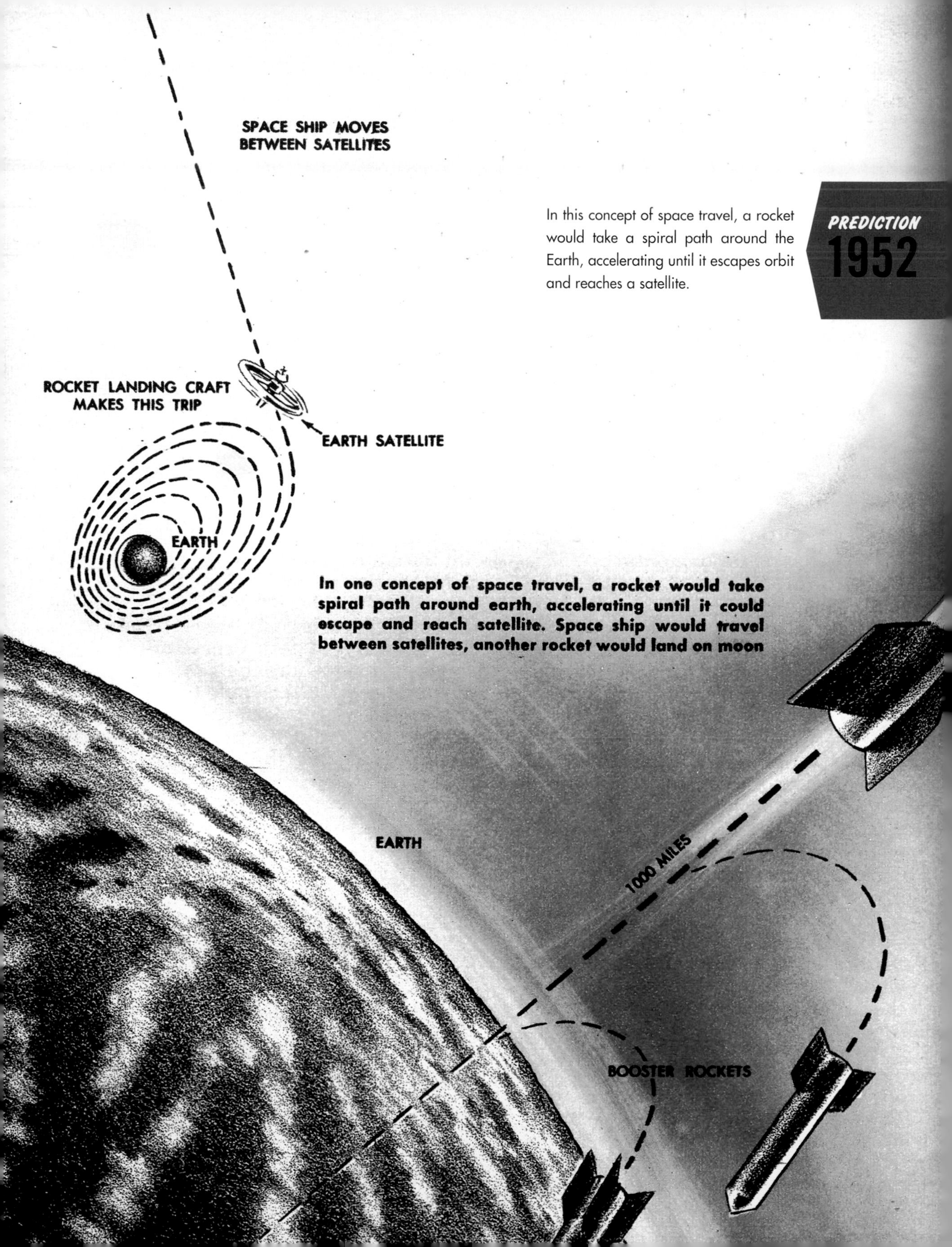

PREDICTION
1952

In this concept of space travel, a rocket would take a spiral path around the Earth, accelerating until it escapes orbit and reaches a satellite.

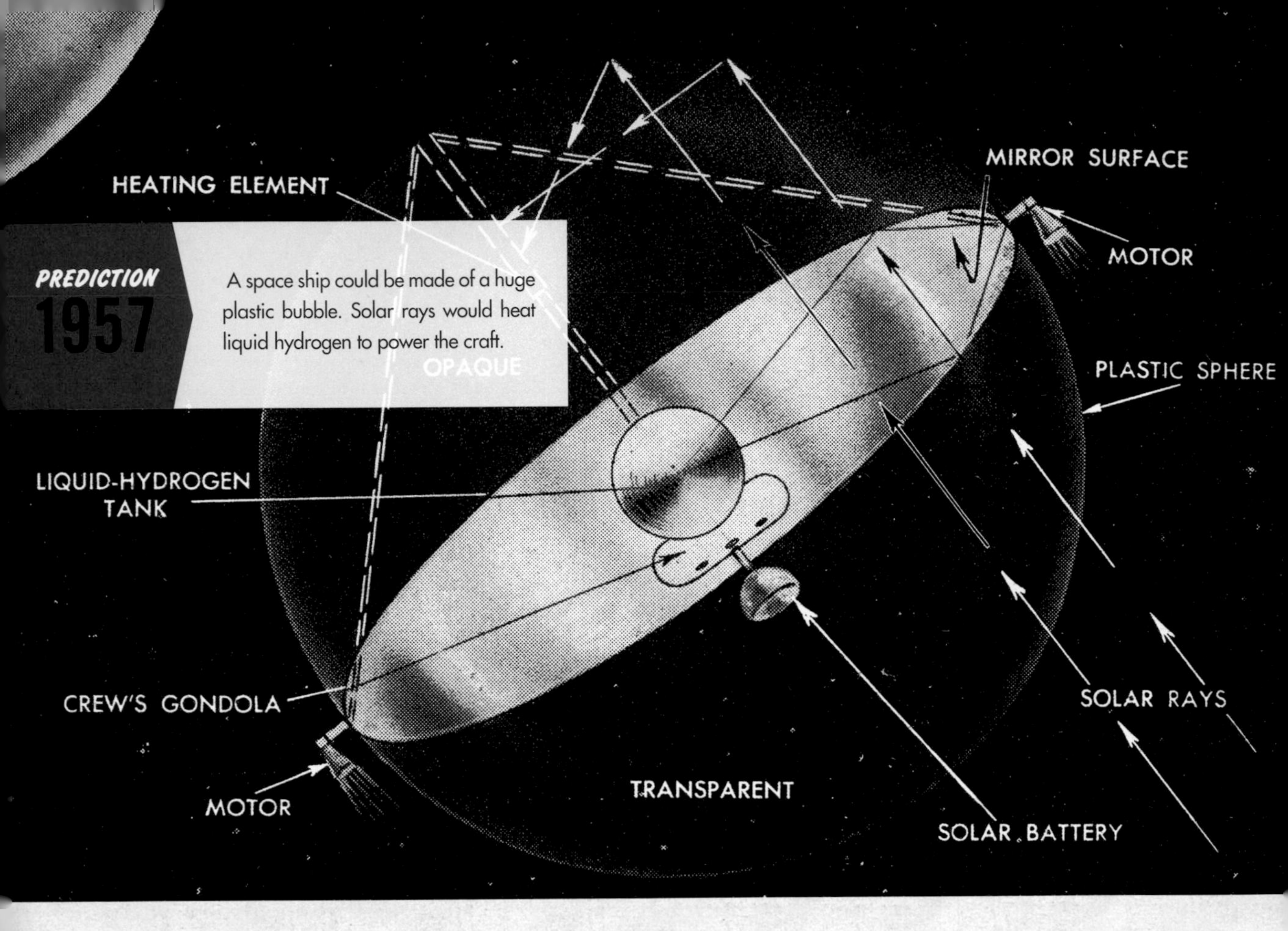

PREDICTION 1957 A space ship could be made of a huge plastic bubble. Solar rays would heat liquid hydrogen to power the craft.

HAPPIER AND HEALTHIER

PREDICTION 1939 The city of tomorrow—Democracity—will be a central town or "brain center" surrounded by satellite towns, all containing factories.

People who work in these factory towns are going to live close to their work. They will have school and a movie. The resident of the satellite town will get up in the morning in a house pleasant surrounded by green. He can walk to work and his children can walk to school in complete safety because they will never cross a vehicular street. When his wife goes to market, she can walk if she wishes, through a park, and she will shop in a park, since the stores will be situated around pleasant green belts. She might drive, but she never would run the risk of killing anyone because no one will cross the highways.

There will be a public market where the farmers living in the green belt will bring their produce. People will not eat concentrated food capsules; they will eat fresh green foods direct from the gardens. They won't have artificial flowers because they will get fresh ones from the garden. The family will develop good taste because they will be surrounded by good things—music, trees and other cultural advantages.

A man will be loyal to his employer because he makes this city possible. Today a man leads two lives: the life of his work and the life of his family. Tomorrow's city is going to inject a third stage, which is playtime to improve his mind and body. The idea of haste will be removed. Today we are rushing to do something. We will temper all that with leisure, therefore nervous diseases will be eliminated to a large extent. We will be happier and healthier because of our surroundings.

PREDICTION

No more bouts with the razor for the man of tomorrow. He'll whisk away whiskers with a chemical solution.

New clothes, rather than new clothing, will be designed for the future. We will accept the human body and make every attempt to bring out its good points. The man will plug his suit into an electric-light socket, then set the meter at the temperature which will keep him comfortable all day, either hot or cool days. The suit will have properties to retain that temperature.

In this future city you need not be exposed to inclement weather, but clothes will be provided for people who want to walk in the rain because of the health-giving properties. Clothes will be designed as the city is designed—for the convenience of the people in it—and they will wear clothes in which they will be comfortable, thus creating style around comfort, convenience and health.

We see no great amount of crime in tomorrow's city because everybody is so happy, thus we will have a police force of only 100 men. Slums develop criminals, but we will have no slums. Instead of policemen we will have service monitors whose functions will be to assist the citizens rather than to correct them in the parking of cars. There will be only a small fire-fighting force because our city will be fireproof and because you will be able to extinguish flames by pushing a button that controls an acid-shooting apparatus.

Drawing of transportati[illegible] system suggested by t[illegible] author. Above, six-la[illegible] system in silhouette. Le[illegible] future metropolis as d[illegible] signed by Mr. Teague f[illegible] U. S. Steel exhibit at Ne[illegible] York World's Fair

PREDICTION 1940

People relax while enjoying the vista of tomorrow's metropolis.

The six-lane transportation system of the future.

PREDICTION 1940 A better world than we have ever known can and will be built. But the builders must feel into the future, groping carefully, ready to adapt their plan, at any moment, to new truths and unforeseen conditions that may be revealed as they progress.

Our better world may be expected to make equally available for everybody such rare things as interesting, stimulating work, emancipation from drudgery and a gracious setting or daily life, freedom of movement, free exchange of thought, bodily well-being and mental equanimity. But since even such conveniences as modern kitchens and bathrooms have not yet become general in America, attainment of these more difficult objectives by a majority of our people is far in the future.

PREDICTION 1950 In the year 2000, any marked departure from what your fellow citizens wear and eat and how they amuse themselves will arouse comment. If old Mrs. Underwood, who was born in 1920, insists on sleeping under an old-fashioned comforter instead of an aerogel blanket of glass puffed with air so that it is light as thistledown, she must expect people to talk about her "queerness." It is astonishing how easily the great majority of us fall into step with our neighbors. And after all, is the standardization of life to be deplored if we can have a standardized helicopter, luxurious standardized household appointments, and food that was out of the reach of any Roman Emperor?

PREDICTION 1939 TOP, Democra-city's opera house has a reversible stage facing open-air amphitheater in the summer and the interior of the theater in the winter. BELOW, a model of an open-air restaurant.

INDEX OF PREDICTIONS BY YEAR

1903, car of the future, 140
1905
cold air health treatment, 109–111
controlling weather, 174
electric farm of the future, 174
electric handshake, 96
jet-powered air-supported trains, 132–134
mail-sorting machines, 82–86
1906, 1,000-ft. steamers, 128
1907
city of the future, 23
flying machines, 156
gyroscopes and monorails, 135
monorails, 134–136
skyscrapers, 23
1909
aeroplanes, 154
internal-combustion battleship engines, 128–131
1910, ironing cracks from damaged furniture, 55
1911
electric energy, 174
fog dispellers, 175
1912, motor-sleigh, 12
1913
city of the future, 27
clothing and accessories, 50–54
fireproof clothing, 50
phototropic-colored clothing, 54
skyscrapers, 27
1914
air-propelled single-tire vehicle, 142
colors to lift spirits, 180
1915, colors reflecting emotions, 181
1918
home of tomorrow, 58
superheated air system, 58
1921, mail delivery by parachute, 83
1922
airplanes, 154, 158–159
home of tomorrow, 38–39, 42
prefabricated $2,000 home, 42
roll-up walls, 38
1923
city of the future, 22
energy from air, 188
radio motion pictures, 90
1924
city of the future, 22–27
Death Rays, 186
light music, 182, 183
radio technology, 87–89
receding, terraced architecture, 22
skyscrapers, 27
surgery on conscious patients, 111
1925
sea exploration, 173
tuberculosis elimination, 112–113
ultraviolet health treatment, 110
1926
airplane skyscraper landing platform, 187
facsimile transmissions, 91
frozen food organization, 62
home of tomorrow, 28, 62
radio technology, 89
synthetic food from coal, 66
tailless airplane, 127
1927
colonizing other planets, 193
end of age of steel, 28
1928
airplane-automobile, 149
airplane landing fields, 24
airplane mail delivery and pickup, 79
city of the future, 18, 20, 24–26, 27
elevated sidewalks, 25
eyes adapt to artificial light, 115
flying ambulance, 107
Giant Transatlantic Flying Boat, 157
home of tomorrow, 41, 42, 45, 58, 62
hornlike wall construction, 45
monorails, 136
obsolescence of milk bottle, 62
pneumatic chairs and modular furniture, 59
receding architecture, 20
rocket planes, 194–195
rooftop lakes for air conditioning, 47
skyscrapers, 25, 26, 27
underground traffic, 24
watercraft advancements, 130
1929
airship of the future, 191
aluminum clothing, 54
asbestos dresses, 54
casein-fiber fabrics, 51
insulating against gravity, 190
1930
airliner of the future, 123
city of the future, 19
revolving restaurants, 19
space travel, 162
1931
city of the future, 32
monorails, 133
soundproof buildings, 32
suburbs and infrastructure, 32
1932
air-propelled glider boats, 129
automobile exhaust and gas masks, 30–31, 32
breeding/growing humans, 115
city of the future, 27, 30–31, 32–33
dashboard stop-and-go lights, 143
science and technology, 8
skyscrapers, 27
wireless phones and televisions, 32–33
1933
plastic automobile bodies, 28
plastic- or synthetic-encased bikes, 49
1935
home of tomorrow (automobile trailers), 43
smell prints, 185
1936
city of the future, 29
glass city, 29
1937
cooking with radio waves, 62
frozen foods, 67
heavy water extending life, 116
home of tomorrow, 45, 62, 67
plastic/synthetic homes, 45
whole-body hearing, 117
1938
dust-free air ("dust magnet"), 54–55
interplanetary rocketport, 170
light concerts, 182
radio-delivered facsimile newspapers, 89–90
removing/replanting diseased organs, 113
traffic balloons, 80
1939
city of the future (Democra-city), 33, 198-199, 202
electric home of tomorrow, 74, 75
exercise on Mars, 172
extended lifespan, 116
home control centers, 75
"mechanical nose," 186
"mood control" lighting, 184
radio-controlled farm, 87
radio living room of tomorrow, 91
1940
car of the future, 140–143, 145
future world, 202
grass in foods, 67–70
high-speed torpedo-like rail cars, 137
high-tech grocery shopping, 71
six-lane transportation system, 200–201
stronger glass, 28
video telephones, 98
1942
"aerogel" bubble construction, 178–179
home of tomorrow, 42–44
push-button phones, 81
super-fast home construction, 42–44
ultraviolet-light air purification, 55
1943
aerocars, 149–150
Helicab, 151

1944
city of the future, 32
completely air-conditioned homes, 61
facsimile transmissions, 95–98
helicopter "airbus," 153
home of tomorrow, 32, 61, 64
kitchen fixtures, 64
motion picture production and transmission, 91–95
projection television, 92–94
underground pneumatic waste tubes, 32
1945
clothing washer technology, 61
hot-air cooking, 61
rocket plane, 190, 192
1946
colorful plastic fabrics, 53
home of tomorrow, 45–48
jet propelled ocean liners, 131
plastics, 45–48
1947
frozen (TV) dinners, 65
space travel, 193, 194-196
vein banks, 113
1948
electronic "sound-readers" for blind, 117
ultrasonic washing technology, 60
1949
cow-milking apparatus, 66
uranium prospecting, 104
1950
aerial transportation, 33
car of the future, 150–151
chemical hair removal, 199
city of the future, 33–35
curing incurable diseases, 114
extended lifespan and youthfulness, 116
facsimile transmissions, 90
frozen and "synthetic" foods, 70
home of tomorrow, 48, 54, 55–57
hosing down home furnishings, 55–57
hypothetical suburb (Tottenville), 33–35
metallurgical advances, 48
petroleum-based clothing, 54
solar power and sootless city, 33, 34–35
standardization of life, 202
supersonic planes, 159
transcontinental railroad tunnel, 136–138
weather prediction instruments, 175
1951
atomic aircraft, 154–160
personal helicopters, 124–125
space station, 196
space travel, 163
1952
artificial heart/lung machine, 118
microwave cooking, 64
reconstructing solar system, 196
1953
dome pillows covering stadiums, 178, 179–180
nuclear-powered aircraft, 161–162
supersonic planes, 160–162
1954
automated language translation, 98–99
computer technology, 98–99
wall-mounted television, 95
1955
anti-aging radiation, 115
mail-sorting machines, 84–85, 86
1956
high-speed bus, 144
home of tomorrow, 40, 64
kitchen appliances, 64
"Picture-Phone," 40
1957
automated ideal home environment, 56
computers diagnosing illness, 108
"flying fan" vehicles, 148–149, 151–154
helicopter-airplane combination, 160
home of tomorrow, 29, 56
personal aerial vehicle, 151–154
pneumatic-tube highways, 146
shelters of sprayed plastic, 29
space travel, 198
transcending limits of senses, 184
1958
interpreting telephone, 99
Project Pied Piper (Big Brother), 162
satellites, 164, 165
satellite-tracking camera, 165
space travel and space stations, 162–167
1959
fusion energy, 186–188, 189
information-retrieval systems, 99–101
lunar exploration, 166
new types of penicillin, 114
1960, nuclear-tipped-missile submarines, 126
1961
personal aerial vehicle, 152–153
space walks and space suits, 167
1962
cellulose foods, 70
computer technology and pocket computers, 101
1963
automated dining table prep and clean-up, 64
home of tomorrow, 58–61, 64
induction cooking without heating range top, 58–61
knocking out hurricanes, 176–177
1964
biochemical pacemakers, 113
personal rocket-packs, 192–193
1965
"Autoline" automated car travel, 143
hydrofoils, 132
jet-powered automated "tube trains," 138–140
supersonic airliner, 161
ultrasound technology, 111–112
1966
car of the future, 144
computers and medicine, 108
radio-wave electricity, 188
sleep-inducing machines, 117–118
tooth transplants, 114
1967
automated car travel, 147
car of the future (airplane-type controls), 144–145
computer-run home, 70–75
heart booster, 119
lunar exploration, 167
1968
airport shuttle alternatives, 139
lake-bottom airport, 155
laser "knife" scalpel, 111
monorails, 139
"people sniffers" and olfactronics, 184, 185–186
"seeing-eye" canes, 118
wristwatch communication centers, 100
1969
bathroom of the future, 63
diamond surgical blades, 109
1970, breeding/growing humans, 116

INDEX

A

Aerocars. *See* Air travel
"Aerogel" bubble construction, 178–179
Air conditioning
 in homes/buildings, 40, 46–47, 61
 in vehicles, 141, 144
Airplanes. *See also* Air travel
 atomic-powered, 154–160
 delivering/picking up mail, 78, 79, 90
 early-1900s predictions, 154, 156–158
 evolution overview, 122
 flying ambulance, 107
 helicopter-airplane combination, 160
 motorless glider, 158–159
 radios guiding, 88
 supersonic, 159, 160–162
 tailless, for fool-proof flying, 127
Airport shuttle alternatives, 139
Air purification, 54–55
Air travel, 120–167. *See also* Airplanes; Space exploration
 airliner and airship of the future, 123, 191
 "flying fan" vehicles, 148–149, 151–154
 Giant Transatlantic Flying Boat, 157
 helicopter "airbus," 153
 helicopter-airplane combination, 160
 lake-bottom airport, 155
 nuclear-powered aircraft, 161–162
 overview of predictions and reality, 122–125
 personal aerial vehicles, 122, 148–154
 personal helicopters, 124–125, 150–151
 personal rocket-packs, 192–193
 rocket planes, 159, 190, 194–195
 skyscraper landing platform, 187
Aluminum clothing, 54
Ambulance, flying, 107
Anti-aging radiation, 115
Asbestos dresses, 54
Automobiles, 140–147
 airplane-like controls, 144–145
 air-propelled single-tire vehicle, 142
 automated travel in, 143, 147
 with circular lounge in rear, 145
 flying. *See* Air travel, personal aerial vehicles
 Jeane Dixon on, 144
 plastic replacing metal in, 28–29, 140–143
 in pneumatic-tube highways, 146
 purifying exhaust of, 31
 rear-engine design, 141
 safety features, 143, 144
 six-lane transportation system and, 200–201
 stop-and-go lights on dashboard, 143
 trailers pulled by, 43
 transcontinental trips, 140

B

Balloons, for traffic, 80
Bathroom, as activity center, 63
Bernal, J. D., 10, 11, 15
Bikes, encased in plastic, 49
Biochemical pacemakers, 113
Blind, "sound-readers" for, 117
Braun, Dr. Wernher von, 196
Brennan, Louis, 134–136
Bridges, 21, 22, 25
Brown, Edward F., 32
Bus of tomorrow, 144
Bus-trains, 139
Butler, Frank Hedges, 154

C

Casein fibers, 51
Cities of the future, 16–35
 age of steel ending, 28
 automobile exhaust and gas masks, 30–31, 32
 bridges, 21, 22, 25
 countryside indistinguishable from, 32–35
 elevated sidewalks and sunken streets, 22–27
 happier and healthier in, 198-199
 overview of predictions and reality, 18–21
 revolving restaurants, 19
 skyscrapers and building architecture, 18, 19, 20, 21, 22–27, 29
 sootless garden city, 32
 soundproof buildings, 32
 suburbs and, 32–35
 underground pneumatic waste tubes, 32
Clarke, Arthur C., 13, 14, 78, 81
Clavilux, 182
Cockerell, Christopher, 138
Cold air health treatment, 109–111
Colors to lift spirits, 180–184
Communication, 76–101. *See also* Mail; Radio; Telephone; Television
 facsimile transmissions, 89–90, 91, 94, 95–98
 overview of predictions and reality, 78–81
Computers
 controlling highway traffic, 143
 ECHO (Electronic Computing Home Operator) systems, 70–75
 forecasting weather, 175
 functions of, 41
 in home, 70–75
 initial predictions about, 13
 looking up library references, 99–101
 in medicine, 104, 108
 pocket-size, 101
 synthetic speech from, 101
 translating Russian to English, 98–99
 in watches, 100
Countryside, like cities, 32–35

D

Daylight glasses, 180
Death Rays, 186
de Ferranti, S Z, 174
de Forest, Dr. Lee, 64, 89, 91
Democra-city, 33, 198-199, 202
Diamond surgical blades, 109
Dixon, Jeane, 144
Dust control, 54–55, 56
Dyson, Freeman, 14

E

ECHO (Electronic Computing Home Operator) systems, 70–75
Einstein, Albert, 10, 190
Electric farms, 174
Electric handshake, 96
Electric home, 70–75
Eyes adapting to artificial light, 115

F

Fabrics, 50–54
Facsimile transmissions, 89–90, 91, 94, 95–98
Falk, Roland, 161–162

Farms
electric, 174
radio-controlled, 87
Fashion and fabrics, 50–54
aluminum clothing, 54
asbestos dresses, 54
casein fibers, 51
colorful weaves, 53
fireproof clothing, 50, 54
plastic fabric, clothing, and accessories, 50, 53, 54
Foa, Dr. Joseph, 139–140
Fog dispellers, 175
Food
automated shopping for, 71
freezing, to maintain freshness, 67
grass in, 67–70
from sawdust and wood pulp (cellulose), 70
synthetic, from coal, 66
Frozen foods, 65, 67
Fuller, Buckminster, 18, 44
Furniture
hosing down, 55–57
ironing cracks from, 55
modular, pneumatic chairs and, 59
plastic, 56, 61
Fusion energy, 186–188, 189, 196
Future
imagination and, 14, 15, 172
twenty-first century, 14, 21, 171–172

G

Glass
bottles, obsolescence of, 62
city of, 29
stronger, 28
Grass in food, 67–70
Grindell-Matthews, H., 186
Grocery "assembly line," 71
Gyroscopes and monorails, 135

H

Hafstad, Doctor, 160
Hair removal, with chemicals, 199
Handshake, electric, 96
Health and medicine, 102–119
anti-aging radiation, 115
biochemical pacemakers, 113
breeding/growing humans, 115, 116
cold air treatments, 109–111
colors to lift spirits, 180–184
computers and, 104, 108
curing incurable diseases, 114
diagnostic techniques, 108, 112
diamond surgical blades, 109
electronic "sound-readers" for blind, 117
expected life spans, 104–106, 116
extending life, 115, 116
eyes adapting to artificial light, 115
flying ambulance, 107
heart booster, 119
heart/lung machine, 118
heavy water extending life, 116
new types of penicillin, 114
overview of predictions and reality, 104–107
removing/replanting diseased organs, 113
"seeing-eye" canes, 118
sleep-inducing machines, 117–118
surgery on conscious patients, 111
tooth transplants, 114
tuberculosis elimination, 112–113
ultrasound technology, 111–112
vein banks, 113
Heavy water extending life, 116
Helicopters. *See* Air travel
Home of tomorrow, 36–75. *See also* Food; Kitchen
"aerogel" bubble construction, 178–179
air conditioning, 46–47, 61
automated dining table prep and clean-up, 56, 64
automobile trailers, 43
bathroom as activity center, 63
computer home operator system, 70–75
dust control, 54–55, 56
electric home, 70–75
fashion and, 50–54
hornlike wall construction, 45
hosing down interior furnishings, 55–57
housekeeping and cleaning, 54–57
ironing cracks from damaged furniture, 55
metallurgical advances affecting, 48
"mood control" lighting, 184
overview of predictions and reality, 39–41
perpetual sunshine, 42
plastic/synthetic materials, 45–48
pneumatic chairs and modular furniture, 59
prefabricated $2,000 home, 42
proof against fire, vermin, weather, 45–48
roll-up walls, 38
rooftop lakes for air conditioning, 46–47
super-fast construction, 42–44
superheated air system, 58
ultrasonic washing technology, 60
ultraviolet-light air purification, 55
washing clothes, 60–61
Housekeeping and cleaning, 54–57
Hurricanes, knocking out, 176–177
Hydrofoil liner, 132
Hydroplane, 130

J

Jeep-of-the air, 153–154
Jefferson, Thomas, 10

K

Kitchen. *See also* Food
appliances, 64
cooking with radio waves, 62
dishwashers, 64
frozen food organization, 62
frozen (TV) dinners, 65
induction cooking without heating range top, 58–61
microwave cooking, 64, 75
obsolescence of milk bottle, 62
solar cooking, 35

L

Lang, Fritz, 21
Langmuir, Dr. Irving, 138
Life spans, 104–106

M

Mail
airplane delivery and pickup, 78, 79, 90
parachute delivery, 83
sorting machines, 82–86
"Mechanical nose," 186
Medicine. *See* Health and medicine
Metallurgical advances, 48
Microwave cooking, 64, 75
Milking apparatus, 66
Monorails, 133, 134–136, 139
"Mood control" lighting, 184

N

Nelson, Paul, 45

O

Opel, Fritz von, 194

P

Pacemakers, biochemical, 113
Parachute mail delivery, 83
Penicillin, new types of, 114
"People sniffers" and olfactronics, 184, 185–186
Personal aerial vehicles, 122, 148–154
Personal helicopters, 124–125, 150–151
Personal rocket-packs, 192–193

Plastic
"aerogel" bubble construction, 178–179
buttons, 50
clothing, fabric and accessories, 50, 53
dome pillows covering stadiums, 178, 179–180
encasing bikes, 49
furniture, 56, 61
homes, 45–48
proliferation of, 171
replacing metal in cars, 28–29, 48, 140–143
sprayed, shelters of, 29
Pneumatic technology
furniture, 41, 59
highways, 146
mail-sorting machines, 82–86
underground waste tubes, 32
Predictions. *See also specific topics*
accuracy of, 10–11, 15
imagination and, 14, 15
nature of, 9–10, 13–14
this book's perspective, 11–12, 14–15
by year. *See Index of Predictions by Year*
Project Pied Piper (Big Brother), 162

R

Radiation, anti-aging effects, 115
Radio, 87–98
airplanes guided by, 88
evolution overview, 78
facsimile transmissions, 89–90, 91, 94, 95–98
gearing up for mass audience, 81, 87
motion pictures in home, 90–95. *See also* Television
presidential campaign and, 87
relay stations, 81, 87
vision of "Father of Radio," 89
Radio-controlled farm, 87
Radio-wave electricity, 188
Revolving restaurants, 19
Rocket planes, 159, 190, 194–195

S

Satellites, 78, 81, 147, 164, 165–166, 197. *See also* Space exploration
Satellite-tracking camera, 165
Sea exploration and transportation
air-propelled glider boats, 129
Giant Transatlantic Flying Boat, 157
hydrofoil liner, 132
hydroplane, 130
hydro-powered ocean liner, 130
internal-combustion battleship engines, 128–131
jet propelled ocean liner, 131
new frontier of, 173
nuclear-tipped-missile submarines, 126
1,000-ft. steamers, 128
"Seeing-eye" canes, 118
Senses
"mechanical nose" and, 186
"people sniffers" and olfactronics, 184, 185–186
smell prints and, 185
transcending limits of, 184
Sidewalks, elevated, 22, 25
Six-lane transportation system, 200–201
Skyscrapers, 18, 19, 20, 21, 22–27, 29, 187
Sleep-inducing machines, 117–118
Smell prints, 185
Solar energy, 33, 35, 180, 198
Soundproof buildings, 32
"Sound-readers" for blind, 117
Space exploration, 162–167
atomic-powered rocket, 162–165
colonizing other planets, 193–197
exercise on Mars and, 172
friction, craft design and, 165–167
interplanetary rocketport, 171
launching from space island, 165
lunar exploration, 162, 166, 167
Project Pied Piper (Big Brother), 162
reconstructing solar system and, 196
satellites and, 78, 81, 147, 164, 165–166, 197
satellite-tracking camera, 165
space stations and, 125, 165, 166, 167, 196
space walks and space suits, 167
Standardization of life, 202
Steel, 28, 48
Stout, W.B., 149, 150, 151
Suburbs, 32–35
Surgery. *See* Health and medicine
Sutherland, Jim and Ruth, 70–75

T

Technology
driving predictions, 9–10
early distrust of, 10
evolutionary overview, 11–13
high hopes for, 18–21
science fiction and, 14
social change and, 21
transcending limits of senses, 184
Telephone
answering machines, 94
cables for, 94–95
hands-free auto-dial, 64
initial predictions about, 13
interpreting, 99
push-button, 78, 81, 98
video, or vidphones, 34, 40–41, 64, 78, 81, 98
wireless, 32–33
Television
in bathrooms, 63
in bus of tomorrow, 144
coaxial cables for, 95
color, 94
first commercial broadcast, 78
overview of predictions and reality, 78
projection, 92–94
signals over long distances, 91, 94–95, 98
studio of tomorrow, 91
video phones and. *See* Telephone
wall-mounted, 95
wireless, 32–33
Tooth transplants, 114
Tottenville, 33–35
Traffic balloons, 80
Trains
high-speed torpedo-like rail cars, 137
hovertrain, 138–139
jet-powered air-supported, 132–134
jet-powered automated "tube trains," 138–140
monorails, 133, 134–136, 139
six-lane transportation system and, 200–201
transcontinental railroad tunnel, 136–138

U

Ultrasound technology, 111–112
Uranium prospecting, 104

W

Washing clothes, 60–61
Weather
controlling, 174
fog dispellers, 175
knocking out hurricanes, 176–177
predicting, 175
Wilfred, Thomas, 182
Wristwatches, 100–101

HEARST BOOKS
New York
An Imprint of Sterling Publishing
387 Park Avenue South
New York, NY 10016

The illustrations, text, and photographs in this volume previously appeared in *Popular Mechanics* issues between 1903 and 1970.

Project Editor: Rose Fox
Designer: Laura Palese

Photo and Illustration Credits
Associated Technical Services, Ltd: 100
Roswell Brown: 126
S. W. Clatworthy: 41
Monroe Eisenberg: 139, 155
Phil Huy: 119
The Illustrated London News/
Mary Evans Picture Library: 38, 58
Robert C. Korta: 34
The New York Times/Redux: 30-31
Penn State University Archives,
Pennsylvania State University Libraries: 60
Pomerance & Breines: 46
A. M. Richardson: 178

Library of Congress Cataloging-in-Publication Data
Benford, Gregory, 1941-
The wonderful future that never was : flying cars, mail delivery by parachute, and other predictions from the past / Gregory Benford and the editors of Popular mechanics.
p. cm.
At head of title: Popular mechanics
Includes index.
1. Technological forecasting. 2. Inventions--History. 3. Twentieth century--Forecasts. I. Popular mechanics magazine. II. Title. III. Title: Popular mechanics, the wonderful future that never was. IV. Title: Popular mechanics.
T174.B453 2010
609--dc22
2010003998

10 9 8 7 6 5 4 3 2 1

First Paperback Edition 2012
Published by Hearst Books
A division of Sterling Publishing Co., Inc.
387 Park Avenue South, New York, NY 10016

www.popularmechanics.com

For information about custom editions, special sales, premium and corporate purchases, please contact Sterling Special Sales Department at 800-805-5489 or specialsales@sterlingpublishing.com.

Distributed in Canada by Sterling Publishing
c/o Canadian Manda Group, 165 Dufferin Street
Toronto, Ontario, Canada M6K 3H6

Distributed in Australia by Capricorn Link (Australia) Pty. Ltd.
P.O. Box 704, Windsor, NSW 2756 Australia

Manufactured in China

Sterling ISBN 978-1-58816-975-4